777 Mathematical Conversation Starters

© 2002 by

The Mathematical Association of America (Incorporated)

Library of Congress Catalog Card Number 2002107969

ISBN 0-88385-540-2

Printed in the United States of America

Current printing (last digit):

10 9 8 7 6 5 4 3 2 1

777 Mathematical Conversation Starters

John de Pillis

Published and Distributed by
THE MATHEMATICAL ASSOCIATION OF AMERICA

SPECTRUM SERIES

The Spectrum Series of the Mathematical Association of America was so named to reflect its purpose: to publish a broad range of books including biographies, accessible expositions of old or new mathematical ideas, reprints and revisions of excellent out-of-print books, popular works, and other monographs of high interest that will appeal to a broad range of readers, including students and teachers of mathematics, mathematical amateurs, and researchers.

Polyominoes, by George Martin

Power Play, by Edward J. Barbeau

The Random Walks of George Pólya, by Gerald L. Alexanderson

The Search for E. T. Bell, also known as John Taine, by Constance Reid

Shaping Space, edited by Marjorie Senechal and George Fleck

Student Research Projects in Calculus, by Marcus Cohen, Arthur Knoebel, Edward D. Gaughan, Douglas S. Kurtz, and David Pengelley

Symmetry, by Hans Walser. Translated from the original German by Peter Hilton, with the assistance of Jean Pedersen.

The Trisectors, by Underwood Dudley

Twenty Years Before the Blackboard, by Michael Stueben with Diane Sandford

The Words of Mathematics, by Steven Schwartzman

To order MAA publications, contact:

MAA Service Center
P. O. Box 91112
Washington, DC 20090-1112
800-331-1622   FAX 301-206-9789
www.maa.org

Preface

FOR WHOM IS THIS BOOK INTENDED?

This book shows that there are very few degrees of separation between mathematics and topics that provoke good conversation such as, why language matters, how we know what we know, the value of fame, the objectivity of truth, when fiction trumps truth, the inevitability of progress, the anatomy of thought, and the limits of reason.

Therefore, if you enjoy good conversation, whether you are a mathematician or otherwise, this book is for you.

How To Use This Book

To locate a Conversation Starter or item, you can

a. Drop the book on the table, see where it opens, choose an item (indicated by numbers in square brackets, for example, item [212]), and start from there. Repeat if desired.

b. Consult the Table of Contents or the Index and choose any item.

c. Read the book in the standard way—linearly. Start from page 1 and go on from there.

d. Go to the section "Sampler," on page 1, and choose themes and supporting items presented there.

ACKNOWLEDGEMENTS

This work is rooted in *Mathematical Maxims and Minims*, a collection of quotes published by John de Pillis and Nick J. Rose in 1988.

Many have given of their good time and efforts toward this work. I am pleased to express my gratitude to:

John Baez (University of California, Riverside),

Art Benjamin (Harvey Mudd College),

Tony Chan (University of California, Los Angeles),

Richard Cottle (Stanford University),

Carl de Boor (University of Wisconsin),

Chandler Davis (University of Toronto),

Keith Devlin (Center for the Study of Language and Information),

Allan Edelson (University of California, Davis),

Martin Gardner (science writer),

Gene Golub (Stanford University),

Ron Graham (University of California, San Diego),

Agnes M. Kalemaris (SUNY, Farmingdale),

Al Kelley (University of California, Santa Cruz),

Joshua Lederberg (Rockefeller University),

Tom Lehrer (University of California, Santa Cruz),

Jay Leno (National Broadcasting Co.),

Ken Millett (University of California, Santa Barbara),

Stan Osher (University of California, Los Angeles),

Robert Osserman (Mathematics Sciences Research Institute (MSRI)),

Helga Pollock (National Broadcasting Co.),

Mike Raugh (Harvey Mudd College),

Jean E. Rubin (Purdue University),

Eric Schechter (Vanderbilt University),

Seattle Earthquake of 2001,

Francis Su (Harvey Mudd College).

Although given the option of refusal, all contributors of original quotes were gracious in their willingness to share their space with my cartoons, all of which are original for this book. For their kindness and good sportsmanship, my deepest gratitude.

No book assembles itself—it takes hours of effort by editors and production people. Therefore, I wish to express my appreciation of the professionalism of Elaine Pedreira and Beverly Ruedi of the Mathematical Association of America. And what pleasure it was to work with Professor Gerald Alexanderson whose editorial skill and sound judgment enhanced this work immeasurably.

And to my wife, Susie, our three daughters, Emmeline (University of Hawaii, Hilo), Gretchen, and especially Lisette (Harvey Mudd College) who kick-started this project in the first place, my unbounded affection, love, and thanks.

Legend

 The little man with the boater or pork-pie hat (left of this paragraph) is a marker the separates this author's comments from quotes of other authors.

—JdP

★ *All quotes herein, attributable to the cited author and appearing for the first time in print, are indicated with the star symbol.*

John de Pillis
Riverside, California
jdp@math.ucr.edu

Contents

Contents

Contents

Sampler

CIVILITY

[1]   Civility (which is required in good conversation) implies that all parties listen and that all parties are allowed to speak.

But civility is not a natural state, it is acquired. Therefore, skills of conversation are difficult to learn and to exercise especially in the face of disagreement. When confronted with an unsettling idea or opinion, it is *challenging* to offer either a better idea or an objective rebuttal.

WHAT IS BAD CONVERSATION?

[2]   Lack of civility in conversation manifests itself in several ways. Here are two of them.

1. **Mind-Reading/Motive Analysis.** You are mind-reading when you accuse your conversation partner of having (usually) bad motives or a hidden agenda. Granted, you may occasionally be *correct* in your description of such private, unspoken thoughts—but mind-reading is neither reliable nor is it a compelling, convincing rebuttal.

Distinguish between the idea and the person

   You can be sure you have fallen into the mind-reading trap if, instead of responding to the idea in a discussion, you hear yourself using the word, "you," in phrases such as, "You are trying to. . ." or "You think that. . ."

2. **Attack the person, not the idea (*ad hominem* attack).** It is natural to believe that a bad idea must come from a bad person. This might account for our inclination to respond to a "bad" idea with a personal, or *ad hominem* attack. It is difficult, but necessary, to distinguish between the idea and the person espousing the idea.
   You can be sure you have fallen into the *ad hominem* trap if you indulge in name-calling, or use phrases such as, "anyone who thinks that, must be a [bad thing]," or "if you really believe that, then you are a [worse thing]."

WHAT IS GOOD CONVERSATION?

[3] **Definition. Good conversation** is an exchange of ideas that can be surprising, refreshing, challenging or affirming. Good conversation, like good drama, may be marked by conflict and contrast—and like good drama, good conversation enriches and educates.

CONVERSATION STARTERS: A SAMPLER

The book contains 777 Conversation Starters, or items, numbered in sequence and indicated in square bracket format,

$$[1], [2], [3], \ldots, [193], \ldots, [777].$$

The following section presents eighteen themes, each of which is supported by one or more Conversation Starters or items.

You Don't Need to Be a Mathematician for These

Theme 1 *Some Thoughts of U.S. Presidents.* United States presidents, Abraham Lincoln and George Washington, had thoughts on topics that are (somewhat) related to mathematics. But there are other less known aspects of these two individuals. See:

- *Washington*, items **[768]**, **[769]**
- *Lincoln*, items **[274]**, **[275]**, **[276]**.

Theme 2 *Is Pleasant Fiction More Important than Truth?* It is natural to romanticize—we tend to replace unpleasant realities with happier, more exciting fictions... but in science? See the following Conversation Starters:

- *Embellishing Collective Memory: Apollo 13*, item **[148]**.
- *Drifting Toward Urban Legend*, item **[65]**.

Theme 3 *Is It Only Scientists Who Should Strive for Clarity?* There are several Conversation Starters dealing with the broad themes:

a. Do poets and non-scientists require *ambiguity*?
b. Are "artistic" writers stifled when they abandon ambiguity in favor of clarity?

c. If we accept ambiguity, do we then encourage and endorse Humpty Dumpty thinkers, that is, those whose words mean whatever they choose? (See item [77] for Mr. Dumpty's actual quote.)

d. Who are the Persuaders among us and what are their "clarity responsibilities?"

For more on these themes, see the following Conversation Starters:

- *Critical Thinking*, item [68].
- *Michael Shermer's baloney detection*, item [72].
- *Definitions for the Non-Scientist*, item [77].
- *Gore Vidal and Other Humpty Dumpty Thinkers: Reality Trumps Satire*, item [78].
- *Thomas Kuhn and the Value of Definition*, item [87].
- *Alan Sokal's Puckish Hoax*, item [91].
- *Three Rebuttals to Alan Sokal's Hoax*, items [94], [95], and [96].
- *Alexander Graham Bell's linguistic blunder*, item [99].
- *Cases of unavoidable ambiguity*, items [100]–[102].
- *Aristotle and an avoidable ambiguity*, item [103].

Theme 4 *Fame without Achievement.*   Since scientist Jonas Salk did something significant and worthwhile, we understand why he is famous. His fame is a result of his achievements. How can the teacher's worthwhile achievements be rewarded (a form of fame) when popular culture (*e.g.*, film) teaches that fame and adulation are goals and not the result of any achievement? See

- *Teaching*, item [711].
- *When fame without achievement is a goal* (comments of film maker Griffin Dunne, item [723], *letter to Dear Abby*, items [723], [724], and *awe of celebrity*, item [726].

Theme 5 *Why the Computer Will Not Go Away.*   What is it about the digital computer that makes it so pervasive and (apparently) necessary in our daily lives? Is it necessary to the degree that we are led to believe? Arguably, the arithmetical computing (the adding, subtracting. multiplying, and dividing) done by the computer is among its least important features. For more, read

- *Extending Collective Memory*, item [145].
- *Inventions with minds of their own: when devices develop in unanticpated directions*, item [146].

Theme 6 *Is There a Spiritual Component in Science?* Are scientists all of one mind when it comes to spirituality? See

- *Faith, Religion, and Sprituality*, item [204].

Theme 7 *Is Progress Inevitable?* Do we have cause to be optimistic about the future? Does history confirm that "progress," once started, never loses momentum? For a perspective on the march of progress, read

- *Progress*, items [580]–[587].
- *Mathematics, the exception*, item [443].

Theme 8 *The Structure of Thought.* Thought has structure that is reflected in natural language. The better we understand the anatomy of thought, the less likely we are to be misled and beguiled by faulty logic from Persuaders who, for example, confuse "*P* implies *Q*" with their look-alike sentences, "*Q* implies *P*." For details and examples, see

- *Anatomy of Thought*, items [20]–[35] concern the relationship between "necessary" and "sufficient." Diagrams represent the logic of implications and their equivalences.

Theme 9 *What Are Deduction and Induction All About?* How can a dry spot in a parking lot after a rain illuminate the difference between induction and deduction? Should Sherlock Holmes really be saying "elementary *induction*, my dear Watson," instead of "elementary *deduction*, my dear Watson"? For more on this point, see

- *Deduction vs. Induction*, item [160].
- *Examples of Induction/Deduction*, items [163], [166]. Item [163]. reveals a published incident in which Sherlock Holmes himself confuses deduction with induction.

Theme 10   *Dimension.*   Why study dimensions higher than three? Does it make sense for us, inhabitants of three dimensions, to direct our attention beyond the third dimension? What does the art of computer graphics have to do with four-dimensional space? See the discussions on these points in

- *Dimension*, items **[175]**–**[177]**.

Theme 11   *Non-Technical Examples Going Beyond Our Senses.*   It can be argued that science is the art of extending our five feeble senses. For example, the microscope and the telescope extend our sense of vision. To get an idea of how we go "outside of ourselves," beyond our own senses, read

- *Analogy Is Not Proof*, item **[339]**.

Theme 12   *Time Dilation of Special Relativity in a (Non-Technical) Nutshell.*   According to the theory of Special Relativity, if you observe someone who is in motion relative you, then that person will age more slowly than you do.

- For an intuitive, ***non-technical*** description of why time is seen to slow down on moving platforms, see item **[627]**.

You Might Want to Be a Mathematician for These

Theme 13   *Banach-Tarski Paradox: An Accessible Proof of a Weak Version*   How can it be that a solid sphere can be partitioned into five pieces so that those pieces, when reassembled, form two solid spheres, each the same size as the original one? The Banach-Tarski Paradox is one of the strangest results in all of mathematics. Here is a much weaker form of this paradox: The unit interval can be partitioned into countably many sets which, when rearranged, cover the whole real line. For an accessible proof of this weaker result, read

- *Axiom of Choice*, item **[50]**.
- *An "easy" version of the Banach-Tarski paradox*, item **[55]**.

Theme 14   *Seven Intuition-Challenging Examples.*   This is a section of the unexpected. You will see how an infinite area (of the Celestial Horn) can be painted with a finite amount of ink. Then learn how the misunderstood

Queen must use calculus to show that she is paying a fair price for her napkin rings. The Monty Hall problem shows that, contrary to conventional wisdom, large numbers can sometimes help our intuition much more than small numbers do. For more, read

- *Intuition: Examples that Challenge*, items **[283]**–**[305]**.

Theme 15   *Reason Is Limited.*   Kurt Gödel guarantees that some true statements will always remain unproven. Logic itself is responsible for this limitation, not our lack of cleverness. Gödel's result was so startling, that the research of David Hilbert himself was stopped in its tracks. Hilbert was trying to "mechanize" (or formalize) mathematics in a deterministic way. For an intuitive discussion of Gödel's result, see

- *Built-in Limitation of Reason: Kurt Gödel's Incompleteness Theorem*, item **[324]**.

Theme 16   *Bayes' Theorem in the Courtroom.*   Probability theory gives insight to problems more important than coin flipping. Read the following item to learn about an actual court case in which the defense lawyer misdefines the true event space.

- *Bayes' Theorem and Specious Reasoning*, items **[427]**, **[428]**.

Theme 17   *Computer Technology Alters Concept of Proof.*   Mathematicians learn that proof requires insight, of course, plus a few basic rules of logic like *modus ponens* and substitution. But no more—the rules have changed. Now we (willingly?) defer to the results of computer programs. For example, in proving the four-color problem, a digital computer was essential to perform the required thousands of searches and comparisons. In light of the machine's success, no known "old fashioned" human-executed proof has appeared. Also, effective encryption requires that prime numbers are verified only to a "high probability" that is validated by test runs on a machine. For more details, read

- *Four Color Problem*, item **[219]**.
- *Do we know what a proof is anymore?* Item **[606]**.

Theme 18   *Pythagorean Theorem Unleashed.*   The familiar, centuries-old Pythagorean Theorem has not lost its vitality. In the following, you will find a non-algebraic proof of the Pythagorean Theorem (there are dozens of proofs), a way to optimize your pizza choices (thanks to Bob Osserman and Tom Lehrer), and, with very little knowledge of physics, to see how Special Relativity makes sense in its assertion that time slows down and and distance shrinks for observed, fast moving objects.

- *Pythagorean theorem*, item **[611]**.

- *Non-algebraic proof*, item **[613]**.

- *The Pizza Connection*, item **[615]**.

- *Breaking the Bonds of Three Dimensions*, item **[617]**.

- *Pythagorean Theorem in Special Relativity*, item **[626]**.

- *The Michelson-Morley "failure,"* items **[645]**–**[654]**.

[4]

The theory of Special Relativity with its dilation of time and shrinking of distance (using the familiar Pythagorean Theorem) provides but one example of an imagination-widening concept in science that is too often presented without the raw emotionalism it validly provokes and deserves.

When we look up at the night sky, and wonder, "What stories are unfolding in the heavens right now up there?" then what answer is possible when "right now," "up there," and "simultaneous" lose their meaning?

Reality tells us that identical clocks cannot always remain synchronized, and time itself slips from our grasp like water. Against such Reality, Fiction finds a mighty and worthy contender. (See items **[626]**–**[642]**.*)*

—JdP

Is the Cow on the Cover Saying "Mew" or "Moo"?

[5]

In a private communication from Donald J. Mastronarde, Professor of Classics at the University of California, Berkeley:

> *"At some point before the classical period (5th cent.* BCE), *[the vowel in the name of the letter 'μ'] was pronounced 'oo' as in moon, so 'μ' would have been 'moo.' "*
>
> ***Moral:*** *Some cows are more erudite than we are!*

—JdP

A

"Look, Aardwolf, I will NOT rename you 'Aard-hat,' 'Aard-heart,' or anything else. Aardvark here is 1st on the list, you are 2nd, and that's that!"

AARDVARK

[6]   Mathematics is aardvark (hard work).

> —Keith Devlin*, Executive Director,
> Center for the Study of Language and Information,
> Stanford University (and a mathematician who likes to go first).

ABSTRACTION

Note: See also, "Defense of Abstraction," item [334]

[7]   The human brain finds it extremely hard to cope with a new level of abstraction. This is why it was well into the eighteenth century before mathematicians felt comfortable dealing with zero and with negative numbers, and why even today many people cannot accept the square root of minus-one as a genuine number.

But software engineering is all about abstraction. Every single concept, construct, and method is entirely abstract. Of course, it doesn't feel that way to most software engineers. But that's my point. The main benefit they got from the mathematics they learned in school and at university was the experience of rigorous reasoning with purely abstract objects and structures. Moreover, mathematics was the only subject that gave them that experience. It's not what was taught in the mathematics class that was important; it's the fact that it was mathematical. In everyday life, familiarity breeds contempt. But when it comes to learning how to work in a highly abstract realm, familiarity breeds a sense of, well, familiarity—meaning that what once seemed abstract starts to feel concrete, and thus more manageable.

—Keith Devlin, Executive Director,
Center for the Study of Language and Information,
Stanford University Devlin's Angle, MAA Online, Oct 2000,
http://www.maa.org/devlin/devlin_10_00.html

[8]   The modern, and to my mind true, theory is that mathematics is the abstract form of the natural sciences; and that it is valuable as a training of the reasoning powers not because it is abstract, but because it is a representation of actual things.

—T. H. Sanford

[9]   Mathematics is the tool specially suited for dealing with abstract concepts of any kind and there is no limit to its power in this field.

—Paul Dirac (1902–1984)
In Philip J. Davis and Reuben Hersh
The Mathematical Experience, Boston: Birkhäuser, 1981.

Note: See also item [324] for a discussion of Kurt Gödel's Incompleteness Theorem, which proves that there will *always* be true statements that can never be proved.

Compare Lobatchevsky's sentiment (item **[10]** *following)*
with Hardy's famous assertion (item **[416]***) that number*
theory will never have a warlike purpose.

—JdP

[10]   There is no branch of mathematics, however abstract, which may not
some day be applied to phenomena of the real world.

—Nikolai Lobatchevsky (1792–1856)

[11]   No more impressive warning can be given to those who would con-
fine knowledge and research to what is apparently useful, than the reflection
that conic sections were studied for eighteen hundred years merely as an abstract science, without regard to any utility other than to satisfy the craving
for knowledge on the the part of mathematicians, and that then at the end of
this long period of abstract study, they were found to be the necessary key
with which to attain the knowledge of the most important laws of nature.

—Alfred North Whitehead (1861–1947)

Not all mathematicians are in love with abstraction. Read
on for the comments of Russian mathematician, Vladimir
Igorevich Arnold.

—JdP

[12]   Mentally challenged zealots of "abstract mathematics" threw all the geometry (through which connection with physics and reality most often takes place in mathematics) out of teaching.

—V.I. Arnold, Moscow University
*from an address given at a conference
on the teaching of mathematics,
Palais de Découverte, Paris,
7 March 1997.*

(See item [234] for Hilbert's related thoughts on geometry and reality.)

ALGEBRA

 Some students harbor the secret belief that algebra is basically useless and it was only created by mean-spirited teachers who want to increase the misery of all students. (In [283] and [292], however, you will see how algebra performs better than our intuition!) Dave Barry has weighed in on this matter. Here are some of his comments.

—JdP

[13]   [If] co-workers [who are dividing the cost of a restaurant lunch check] know their mathematics, they can easily come up with EXACTLY the correct answer. They can do this using "algebra," which was invented by the ancient Persians. (They also invented the SATs, although they got very low scores because in those days, there were no pencils.) The way algebra works is, if you don't know exactly what a number is, you just call it "X." The Persians found this was a BIG mathematical help in solving problems:

PERSIAN WIFE (suspiciously): How much have you had to drink?
PERSIAN HUSBAND: I had "X" beers.
PERSIAN WIFE: How much is THAT?
PERSIAN HUSBAND: It's a (burp) variable.
PERSIAN WIFE (not wanting to look stupid): Well, OK then.

Historical Footnote: Several years later, when the ancient Romans invented Roman numerals, and it turned out that "X" was actually equal to ten, there was BIG TROUBLE in Persia.

—Dave Barry, humorist,
April 22, 2001

[14]   Algebra is generous; she often gives more than is asked of her.

—Jean d'Alembert (1717–1783)

(To see how algebra "gives more" and takes us to levels of understanding beyond our intuition, see items **[283]**–**[292]**.*)*

[15]   It is well known that every PID is a UFD. But less well known that every IUD is a UFO.

—Lew Lefton, Georgia Institute of Technology

[16]   As long as algebra and geometry traveled separate paths, their advance was slow and their applications limited. But when these two sciences joined company, they drew from each other fresh vitality and then forward marched at a rapid pace toward perfection. It is to Descartes that we owe the application of algebra to geometry—an application which has furnished the key to the greatest discoveries in all branches of mathematics.

—Joseph-Louis Lagrange (1736–1813)
quoted by Morris Kline in *Mathematical Thought from Ancient to Modern Times* (vol. 1), Oxford University Press, p. 322

ANALYTIC GEOMETRY

[17]

Cauchy was asked by a seaman
If functions were made by a demon.
   He said to his critic,
   "You seem analytic,
But, I'd better check it with Riemann."

—Paul Ritger
quoted in *Mathematical Maxims and Minims*, by Nick Rose, 1988.

[18]

"Integrals," as shown by Cauchy,
"Of regular functions of z
   Are really quite dull
   For their value is null
On simple closed curves of B.V."

—James P. Burling
The American Mathematical Monthly, vol. 72, no. 5, 1965, pg. 527.

[19] [Analytic Geometry] immortalized the name of Descartes, and constitutes the greatest single step ever made in the progress of the exact sciences.

—John Stuart Mill (1806–1873)

Note: See also items **[575]**–**[583]** and **[443]** for comments on scientific progress.

ANATOMY OF THOUGHT: THE IMPLICATION

"P Implies Q" = "Sufficient Implies Necessary"

Example: Cordelia's Problem

[20] Cordelia was upset. After we ordered our coffee, Cordelia related her frustration at last night's meeting as she tried in vain to obtain increased funding for the special education classes she was teaching.

Cordelia stirred her coffee with more vigor than usual as she said, "You won't believe how the discussion ended. The director told me outright, 'When you reach my age, Cordelia, you will see that I am right. Case closed!' Well, let me tell you, I know lots of people who have seen many birthdays yet they often show very, very poor judgment."

"When you're right, you're right," Anvil Willie added. "Sure, age often produces a kind of wisdom. But not always. Age alone is not enough."

"Good for you, Willie," Hutch exclaimed, as she placed her hand on his shoulder. "You identified the very essence of Cordelia's frustration—the confusion of 'necessary' with 'sufficient.' "

"Yes, I suppose that is what is disturbing me," Cordelia confirmed. "My director thinks aging is sufficient to guarantee wisdom and I say age is only necessary. There is a difference."

"We can put that piece of logic into pictures," Hutch said as she put her notepad on the table with a flourish. "Let's diagram this classic confusion of 'necessary' with 'sufficient' which, by the way, is the same as confusing 'P implies Q' with 'Q implies P.' "

We could only imagine what kind of picture Hutch would produce. We watched in anticipation as she flipped through the pages of her notepad, searching for a blank sheet. We shoved our coffee cups aside for a better view as she sketched a rough version of the following figure (item **[21]**):

[21]

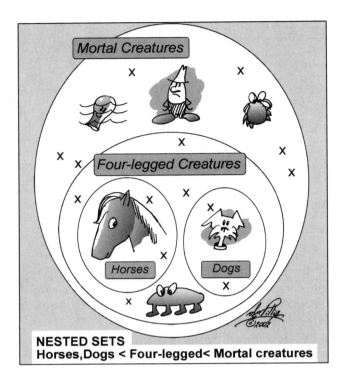

We looked at Hutch's drawing, then at each other. We had to wonder what these horses, dogs and other mortal creatures had to do with Cordelia's problem of whether age implies wisdom.

"But as long as we are talking about implications," Hutch said, "we should first know what the word 'implication' means. That's why I drew this diagram."

She pushed her notepad toward Anvil Willie and invited him to place his finger on an "X" near the picture of the horse and keep it there. This he did with a matter-of-fact shrug.

"OK, Willie, since the X you are pointing to is inside the Horse's oval, your X is a horse."

"Yes, but Willie's X is also inside the larger set (oval) of four-legged creatures," Cordelia exclaimed.

"Bravo, Cordelia," Hutch said. "You have just described the core structure of the implication. Implications are statements that describe one set being contained inside another."

Hutch then wrote the following sentences which she labelled as P, Q, R, and S along with their negatives $\sim P$, $\sim Q$, $\sim R$, and $\sim S$:

[22]

TABLE 1. **Basic Sentences and Negatives for Figure of item [21]**

Basic sentences P, Q, R, S and the negatives	
$P = $ X is a horse	Not $P = \ \sim P = $ X is not a horse
$S = $ X is a dog	Not $S = \ \sim S = $ X is not a dog
$Q = $ X is four-legged	Not $Q = \ \sim Q = $ X is not four-legged
$R = $ X is mortal	Not $R = \ \sim R = $ X is not mortal

"Willie's finger points to horse X," Hutch said, "and pointing to horse X is *sufficient* to produce the *necessary* consequence that Willie is also pointing to the same X as a four-legged creature."

[23] Hutch tabulated Anvil Willie's implication in the following three forms:

TABLE 2. **Mortal Creatures: Three Forms for Anvil Willie's Implication Sentence**

		Sufficient Condition P	$\rightarrow$	*Necessary* Condition Q
1.		X is a horse	implies	X is four-legged
2.	If	X is a horse	then	X is four-legged
3.		P	$\rightarrow$	Q

[24] A "Then" with every "If."   In natural language, Hutch reminded us, a single implication can be written in more ways than shown in item **[23]**. For example, "implies" may be replaced by "therefore." Also, whenever we use the word "if" in a sentence, there is always the implicitly understood word "then," even if it is unwritten. For clarity and completeness, Hutch suggested, within every "if" sentence, there should always be written the word "then."

[25] Linking nested sets with necessary and sufficient.   "Even though your (item **[23]**) gives three different forms for my implication, I notice a common pattern," Anvil Willie interjected. "You always list 'Sufficient' conditions on the left-hand side and 'Necessary' conditions on the right-hand side."

Hutch confirmed Anvil Willie's observation as valid. "Whenever you point to an "X" inside a small set, that is *sufficient* for you to *necessarily* point (to the same X) inside any larger, containing set," Hutch said. She then sketched this summary table:

[26]

TABLE 3. Roles of "Necessary" and "Sufficient" in the
Implication

"*P* implies *Q*" or "*P* → *Q*" or "If *P* then *Q*"
can be restated as
"*P* is *sufficient* for *Q* to be a *necessary* consequence."

More Implication Statements from item [21]. Hutch noted that the natural language of item **[22]** along with item **[21]** can produce other examples of implications. These include the following:

[27]

TABLE 4. Some Implications from item [21])

Mortal Creatures		
Sufficient Condition	→	*Necessary* Condition
X is a horse	implies	X is four-legged
X is a dog	implies	X is four-legged
X is a dog	implies	X is mortal
X is a horse	implies	X is mortal
X is four-legged	implies	X is mortal

[28] "*P* implies *Q*" is not always "*Q* implies *P*" (See item **[36]** for a discussion of special cases where "*P* implies *Q*" is the same as "*Q* implies *P*.")

"Getting back to my original question on age and wisdom," Cordelia said, "I think it can be put into implication format this way." She took Hutch's notepad and wrote:

[29] Cordelia's claim: *P* implies *Q*:

X has Good judgment → X has accumulated many birthdays.

Administrator's claim: *Q* implies *P*:

X has accumulated many birthdays → X has Good judgment.

Cordelia's "*P* implies *Q*" not equal to "*Q* implies *P*"

"That's it, Cordelia," Hutch said. "Your view of what is sufficient (left of "→" in item **[29]**) and what is necessary (right of "→" in item **[29]**) certainly differs from your administrator's view."

"And as we see," Anvil Willie said, pointing to item **[27]**, "You can't switch the order in the implication sentences you have here. If '*P* implies

Q' were always the same as (and as true as) 'Q implies P,' then we would have

'X is a horse implies X is four-legged'

is the same as saying

'X is four-legged implies X is a horse.' "

Hutch slowly sipped her coffee and said, "So we are convinced that 'P implies Q' is not always the same as its converse, 'Q implies P.' But you know," she added with a twinkle, 'P implies Q' *is* equivalent to a very different-looking implication." Turning to a new page of her notepad, Hutch said, "Wanna see?"

[30] "P implies Q" = "$\sim Q$ implies $\sim P$" Always: The Contrapositive

Definition. **The contrapositive** of an implication "P implies Q" is the implication "$\sim Q$ implies $\sim P$," (read "not Q implies not P").

If "P implies Q" is not the same as its converse, "Q implies P," then what *is* it equal to? Another round of coffee was delivered to Table Seven and we settled in to hear Hutch's explanation of the answer that she diagrammed in the following table:

TABLE 5. "P implies Q" = "$\sim Q$ implies $\sim P$"

Two Equivalent Implication Forms
(P implies Q) says the same thing as its contrapositive, namely ($\sim Q$ implies $\sim P$)

A Diagrammatic Demonstration: Hutch promised to show how item **[21]** and item **[22]** validate the claim of item **[30]**. "First, we agree that item **[21]** has already given us the valid statement

[31] P implies Q."

(**Note.** See also item **[23]**.) Hutch opened her notepad to item **[21]** and asked Anvil Willie to place his finger on any "X" outside the set (oval) of "four-legged creatures." With his finger resting on one of the "X's", Hutch made the following two points:

1. Since Anvil Willie's chosen X is NOT a four-legged creature, then, as the picture of item **[21]** shows, that same X, necessarily, automatically, and unavoidably, is NOT a horse, either.

2. In natural language terms of item **[22]**, this says

    [32]   (X is NOT four-legged)   implies   (X is not a horse), or

    $$\sim Q \text{ implies } \sim P$$

Hutch observed that the same diagram (item **[21]**) give us both statements **[31]** and **[32]**, thus establishing the truth of the table of item **[30]**.

EXAMPLES

Anvil Willie and the Monkey's Uncle

We thought for a while. It took some time to feel comfortable with the contrapositive table of item **[30]**. After all, here was a simple rule of human thought that should (intuitively) be complex, not simple.

"The contrapositive is not as exotic and rare as you think, Willie," Hutch said. "For example, you yourself used it the other day when you claimed that the comedy movie we saw together (call it Film [X]) was not funny."

What Anvil Willie said was,

[33]   "If Film [X] is funny, then I'm a monkey's uncle."

Hutch claimed that Anvil Willie was depending on two facts:

1. Our brains are hard-wired to translate his implication to the following contrapositive form,

[34]   "Since I am NOT a monkey's uncle, Film [X] is NOT funny."

2. The conclusion (right-hand side) of **[34]** says Film [X] is NOT funny. This must be accepted as TRUE once we accept the truth of its premise (left-hand side), namely that Anvil Willie is NOT a monkey's uncle. (i.e., TRUE → TRUE)

We all got together to fashion the following implication/contrapositive pairs (item **[35]**) starting with Anvil Willie's film critique, **[33]** and **[34]**.

[35]

TABLE 6. Examples of "*P* implies *Q*" and "~*Q* implies ~*P*"

Implications $P \to Q$ and (equivalent) Contrapositives $\sim Q \to \sim P$	
1. $P \to Q$	(Film X is funny) → (I am a monkey's uncle)
$\sim Q \to \sim P$	(I am NOT a monkey's uncle) → (Film X is NOT funny)
2. $P \to Q$	(X is a dog) → (X has four legs).
$\sim Q \to \sim P$	(X does NOT have four legs) → (X is NOT a dog).
3. $P \to Q$	(X is a live human) → (X breathes oxygen).
$\sim Q \to \sim P$	(X does NOT breathe oxygen) → (X is NOT a live human).
4. $P \to Q$	(X is a good employer) → (X pays a fair wage).
$\sim Q \to \sim P$	(X does NOT pay a fair wage) → (X is NOT a good employer).

We realized that the four special cases in the table of item [35], don't give a rigorous proof that in all cases, sentence "*P* implies *Q*" says the same thing as "~ *Q* implies ~ *P*." But we crafted further examples and that made us more receptive to accepting the equivalence of an implication and its contrapositive.

Anvil Willie was so pleased that his casual statement, item [33], produced such attention and study, that he paid the check for all of us at Table Seven. This was a significant gesture considering we had consumed more than the usual amount of coffee and desserts.

We gathered up our belongings. Our long and pleasant discussion had ended and the sun was beginning to set. We were ready to go home.

—John de Pillis
from *Starlight Café Conversations: An Illustrated
Dictionary from Table Seven.* (See pg. 319.)

[36] "*P* implies *Q*" = "*Q* implies *P*" Sometimes: Equivalence   (See item [28] for a discussion of those cases where "*P* implies *Q*" is not the same as "*Q* implies *P*.")

Sometimes (but not always) the two implications, "*P* implies *Q*" ("*P* → *Q*,") and "*Q* implies *P*" ("*Q* → *P*,"), are both true or both false. This means that their individual "pieces," statement *P* and statement *Q*, are either *both* true or they are *both* false. That is, statements *P* and *Q* are saying the same thing.

To say it another way, one statement (it doesn't matter which) is inferred from the other, and one statement (it doesn't matter which) is deduced from the other.

Examples:

- Assuming the mayor of Hapsberg is determined by a plurality, we have the equivalent statements about Ned:

$$P = \text{Ned is the mayor of Hapsberg.}$$

$$Q = \text{Ned got a plurality of votes in Hapsberg.}$$

P and Q are either *both* true or *both* false. Let's check: If P is false (Ned is not the mayor) then so is Q (Ned did not get a plurality.) If P is true (Ned is the mayor) then so is Q (Ned must have gotten a plurality.)

- Concerning the whole number N, we have the equivalent statements or properties:

$$P = N \text{ is an even number}$$

$$Q = N \text{ is divisible by 2.}$$

Pick your favorite integer N. You will see that P and Q are both true or they are both false. There is no integer for which P is true and Q is false.

- In his book, *Did Adam and Eve Have Navels?* W. W. Norton & Co., 2000, pg. 7, Martin Gardner writes,

> "If Adam and Eve did not have navels, then they were not perfect human beings. On the other hand, if they *had* navels, then the navels would imply a birth they never experienced."

So navels *imply* a birth? Doesn't Gardner mean that from navels we *infer* its cause—namely, that a birth took place?

Actually, Gardner is correct since, in this case, "infer" and "imply" are equivalent. To see this, here are the equivalent statements:

$$P = \text{X has a navel.}$$

$$Q = \text{X was born with an umbilical cord.}$$

We cannot have one of these sentences being true unless the other is true as well. In short, $P \to Q$ and $Q \to P$.

[37] Badly Understood Implications: Poor teachers and unconvincing speakers often misuse the implication "$P \rightarrow Q$."

The strategy is this:

(a) Create a condition P which is obviously TRUE.
(b) Hope the audience accepts the implication "$P \rightarrow Q$" as TRUE. From this, it will follow that the audience must judge condition Q to be TRUE.

Note: Condition Q is always designed to be favorable to the speaker.

[38]

Example 1.   The Foreign Accent

The strategy (in concrete terms):

Hypothesis: P = "X speaks with an accent."

Conclusion: Q = "X is very very intelligent."

Step (a) Establish the truth of hypothesis P by addressing your audience with a thick, foreign accent.

Step (b) Hope the audience accepts the implication $P \rightarrow Q$, which says that use-of-accent implies intelligence.

[39]

Example 2.   Writing, Erasing Quickly

The strategy (in concrete terms):

Hypothesis: P = "X writes and erases quickly."

Conclusion: Q = "X is very very intelligent."

Step (a) Establish the truth of hypothesis P by writing on the board at super-sonic speed and erasing so quickly that your words will survive no more than three seconds.

Step (b) Hope the audience buys into the implication $P \rightarrow Q$, which says that fast writing/erasing implies great intelligence.

[40]

Example 3.   Blocking the Board

The strategy (in concrete terms):

Hypothesis: P = "X blocks the board."

Conclusion: Q = "X is so intelligent, that absent-minded board-blocking behavior necessarily results."

Step (a) Establish the truth of hypothesis P by standing like a post in front of the board.

Step (b) Hope the audience buys into the implication $P \rightarrow Q$, which says that board-blocking is a by-product of an absent-minded genius.

[41] Induction and Deduction in Terms of $P \rightarrow Q$ The symbol "$P \rightarrow Q$" offers a diagrammatic interpretation for understanding the relationship between ***induction (inference) and deduction (conclusion)***.

BLOCK
ANY
CLEAR
VIEW
of the
BLACKBOARD.

Note: See also item **[160]**, "Deduction vs. Inductions".

Recall from items **[26]** and **[27]**, that in writing "$P \to Q$," we have already established that P is a *sufficient condition* to "cause" Q as a *necessary* "consequence." Simply put, left-hand "causes," P, are induced (or inferred) and right-hand "consequences," Q, are deduced. (For a more complete list of terminology used for P and Q in the implication $P \to Q$, see **[162]**, "Popular Synonyms for Inductive and Deductive Reasoning.")

A concrete example illustrating induction and deduction.

Induction of (inferring the) "cause" P: When I see Harry with four legs (Q), I can induce, or infer that he is a horse (P). (We could have induced that Harry is a dog, a cow or some udder animal, so inductions are not as reliable or exact as deductions!)

Deduction of "consequence" Q: When I see Harry is a horse (P), I deduce the necessary "consequence" that Harry has four legs (Q).

Note: Item **[36]** discusses the special case for which induction and deduction coincide. See also "Deduction vs. Induction," item **[160]** and the cartoon, item **[161]**.

ARITHMETIC

[42]   The different branches of Arithmetic—Ambition, Distraction, Uglification, and Derision.

> —Lewis Carroll (1832–1898), pseud. of Charles Lutwidge Dodgson
> *Alice in Wonderland*

[43]

> It used to be fun
> To add one and one
>   But now I'm unsure
>   What sum to secure
> I'm told it may even be none.

> —Leo Moser
> *The American Mathematical Monthly*,
> vol. 80, no. 8, 1973, pg. 902.

[44]   A horse showed an extraordinary ability to learn mathematics. The horse learned arithmetic and algebra and then mastered plane geometry and trigonometry. The horse was now ready for analytic geometry. However, when the subject was put to him, the horse reared up on his hind legs, whinnied, and became very unruly. **Moral:** Never put Descartes before the horse.

[45]   I have no faith in political arithmetic.

—Adam Smith (1723–1790)

ASTRONOMY

(See item **[145]** on the essential role of memory-extending photography in astronomy.)

[46]   Over lunch, Edmond Halley (1656–1742) and Robert Hooke (1635–1703) discussed their shared conviction that the force of gravitation must diminish by the square of the distance across which it is propagated. They felt certain the inverse-square law could explain Kepler's discovery that the

planets move in elliptical orbits, each sweeping out an equal area within its orbit in an equal time.

The trouble was, as Halley noted, that he could not demonstrate the connection mathematically. (Part of the problem was that nobody, except the silent Newton, had realized that the earth's gravitational force could be treated as if it were concentrated at a point at the center of the earth.)

—Timothy Ferris
Coming of Age in the Milky Way,
Doubleday, 1989, pg. 112.

Note: See item **[589]** for a history by Sandro Graffi of the inverse square law in Hellenic science.

[47] An astronomer is one who thinks that if t is sufficiently large, $\sum_1^\infty n!/t^n$ converges and $\sum_1^\infty t^n/n!$ diverges.

—Henri Poincaré (1854–1912)

AUTHOR

[48] Unfortunately what is little recognized is that the most worthwhile scientific books are those in which the author clearly indicates what he does not know; for an author most hurts his readers by concealing difficulties.

—Evariste Galois (1811–1832)

Note: See also Voltaire, item **[112]**, on telling too much.

[49] Next to the originator of a good sentence is the first quoter of it.

—Ralph Waldo Emerson (1803–1882)

AXIOM OF CHOICE

[50]

What the Axiom of Choice Says. Let C be a collection of nonempty sets. Then we can choose a member from each set in that collection. In other words, there exists a function f defined on C with the property that, for each set S in the collection, f(S) ∈ S.

—*JdP*

[51]   At first it seems obvious, but the more you think about it, the stranger the deductions from this axiom seem to become; in the end you cease to understand what is meant by it.

—Bertrand Russell (1872–1970)

[52]

There once was a maiden named Emma,
Who had a peculiar dilemma;
   She had so many beaus,
   That to choose, heaven knows,
She had to appeal to Zorn's Lemma.

—Richard Cleveland
Mathematics Magazine, vol. 52, no. 4, pg. 257, 1979.

[53]

A Glimpse into the Controversy. Since its inception in the early 1900's, the Axiom of Choice has been controversial. In the following paragraphs, Morris Hirsch chronicles some details of this disagreement.

—*JdP*

The fierce debate over the legitimacy of the Axiom of Choice concerned a conflict between two different conceptions of a set: On one side was Frege's logical approach, based on the extension of a concept and the division of everything into two groups according to any kind of rule. On the other side was Cantor's mathematical approach whereby new sets are formed from existing ones according to definite procedures, culminating in Ernst Zermelo's iterative hierarchy of sets. We might call these, respectively, the "top down" and "bottom up" approaches. René Baire (1874–1932), Emile Borel (1871–1956), and Henri Lebesgue (1875–1941), suspicious of arbitrary correspondences, used the "bottom up" approach to treat functions.

Maddy (cited at end of remarks) quotes from a series of letters between these three analysts and their opponent, Hadamard. In 1905, following Zermelo's use of Choice to prove the well-ordering principle, Lebesgue wrote to Hadamard:

[54]

> "The question comes down to this, which is hardly new: Can one prove the existence of a mathematical object without defining it?...I believe that we can only build solidly by granting that it is impossible to demonstrate the existence of an object without defining it."

Who could object to such a reasonable principle? Hadamard could! Admitting that Zermelo had no way of carrying out the mapping needed for a choice function, he insisted that the problem of its effective determination is completely distinct from the question of its existence:

> "The existence ... is a fact like any other."

Today Hadamard's pro-Choice position has prevailed with the vast majority of mathematicians.

—Morris W. Hirsch
Bull. Amer. Math. Soc., Jan. 1995, 32:1, pp. 137–148,
reviewing Penelope Maddy's book, *Realism in Mathematics*,
Oxford University Press, London, 1993.

[55]

Something to sink your teeth into: *In the following discourse, the denisons of Table Seven discuss a seven-step proof of a weak version of the Banach-Tarski paradox, namely, that countably many non-overlapping pieces of a subset of the unit interval, once rearranged, cover the whole real line.*

—*JdP*

"Those math classes have got you to believing almost anything," Anvil Willie exclaimed in disbelief.

"You expect us to believe," Cordelia added, "that you can take a solid ball, cut it up into five pieces, reassemble those pieces and make two solid spheres, each the same size as the original?"

Responding to the challenge, Hutch replied, "Well not in the real world, of course. We can't cut volumes with infinitely many turns. But there is an easier version I could try to show you."

Hutch went on to describe the following example she had learned from her friend, Lisette, a professor at the University:

"I promise you," Hutch exclaimed, "that I can show you a subset of the real interval [0, 1] which is partitioned into countably many sets. I will reorder those countably many sets and, voilà—the whole real line will be covered."

That got our attention and we all accepted Hutch's offer to show us this strange result. First, it was necessary for us to order another round of pastry and coffee. When Hutch took out her pen and placed a large blank pad of paper on the table, here is what happened.

Hutch's "Easy" Example of a Banach-Tarski-like Paradox. Hutch outlined the seven steps, which we promised her, we would try our best to understand.

Step 1. **Define sets B of equivalent numbers.** We throw two numbers x and y into the same set B if and only if $y - x$ is a rational (fraction) q. That is, x and y are in the same set B if and only if, for some rational q, we have

[56]
$$x + q = y$$

In standard terminology, the two numbers, x and y, are said to be *equivalent* to each other. In other words, a number x is thrown into set B along with

[57]

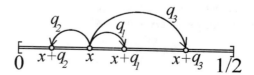

FIGURE 1

All numbers that are a rational (fraction) jump q from x are equivalent to x and to each other.

all its "rational neighbors"—those neighbors y that you can "jump" to from x over a rational distance q. Figure 1, item **[57]** illustrates a number x in $[0, 1/2]$ along with a few of its "rational-distance neighbors," $(x + q_1)$, $(x + q_2)$, and $(x + q_3)$.

Here follow two sets, each of which contains countably many numbers which are separated from each other by a (required) rational distance (jump).

Example 1.

$B_1 =$ all rationals q in $[0, 1/2]$. This is a countable set of numbers (rationals) that are equivalent to each other.

Example 2.

B_2 is the set of all $(\pi/7 + q)$ where q is rational and q is small enough so that $(\pi/7 + q)$ is in the interval $[0, 1/2]$. The three numbers, $(\pi/7 + 1/100)$, $(\pi/7 - 3/100)$, and $(\pi/7)$, are all in the same countable set.

Step 2. **There are uncountably many sets** B_α. The interval $[0, 1/2]$ contains uncountably many numbers. Since each set of equivalent numbers has only countably many numbers, there must be uncountably many sets.

[58]

*WARNING! We are about to use the controversial Axiom of Choice which says we can choose uncountably many numbers, one from each of the uncountably many sets B_α. (See Statement of the Axiom of Choice, item **[50]**.) This axiom generously allows us to make such a choice even though there is no recipe or method that tells us exactly how to do this! For countably many sets, the story is different—a selection process can be described. But for the uncountably many sets, B_α, no such process, or specific recipe, can be described.*

—JdP

Step 3. **Build a core set** A **in** $[0, 1/2]$**, which has uncountably many elements.** Ignoring the warning about the Axiom of Choice, we nonetheless forge ahead and define our core set A to be those uncountably many num-

[59]

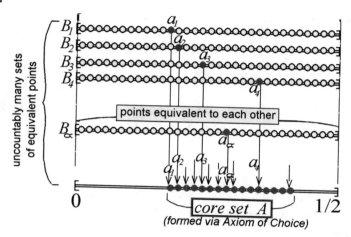

FIGURE 2

bers a_α, one chosen from each of the sets B_α of equivalent numbers (see Figure 2, item **[59]**.)

*No two elements, a and a', chosen by the Axiom of Choice to produce core set A, are equivalent to each other in the sense of equation **[56]**. To see this, suppose the opposite of what we are trying to prove, namely that a and a' are equivalent. (For more on the technique of proof by contradiction, or reductio ad absurdum see item **[598]**) Equivalence of a and a' guarantees that you can jump to point a' from point a by adding a rational number. That is,*

$$\text{for some rational } q, \qquad a + q = a'.$$

*This would mean (again, from item **[56]**) that a and a' belonged to the same set of equivalents B_α. But use of the Axiom of Choice (see statement the Axiom of Choice, item **[50]**) gives us only one representative from each B_α.*

Step 4. Linking a Small Set and a Large Set of Rational Numbers.
Consider the two sets of rational numbers (fractions) as follows:

[60] $$Q_{1/2} = \text{all rationals } q, \text{ where } 0 \le q \le 1/2.$$

$$Q_\infty = \text{all rationals } r.$$

Since sets $Q_{1/2}$ and Q_∞ of rational numbers are each countable, we may list their elements, with subscripts, as follows:

[61] $$Q_{1/2} = \{q_i : 0 \le q_i \le 1/2, \quad i = 1, 2, 3, \ldots\},$$
$$Q_\infty = \{r_i : i = 1, 2, 3, \ldots\}$$

This listing above implicitly gives us a bonus—a one-to-one mapping between elements q_i in $Q_{1/2}$ and r_i in Q_∞ namely,

[62] $$q_i \leftrightarrow r_i, \quad i = 1, 2, 3,$$

where q_i is an element of $[0, 1/2]$, and r_i is an element of $(-\infty, \infty)$.

Step 5. **Defining a "small" set S in [0,1]:** The small set, S, which we will construct, will have countably many non-overlapping pieces. But each piece will be familiar—a copy of (or a shifting of) the core set A. Here we go:

Definition. The "small" set $S \subset [0, 1]$ will consist of all "small" rational shifts of core set A. That is,

$$S = \{A + Q_{1/2}\} \subset [0, 1].$$

More precisely,

$$S = \{A + q_i : q_i \in Q_{1/2} \quad i = 1, 2, 3 \ldots\} \subset [0, 1]$$

Step 6. **Reorder the disjoint pieces of the small set S and thereby cover the real line.** Move each individual piece $A + q_i$ of small set $S_{1/2}$ to a new position on the number line as follows:

$$\text{move} \quad (A + q_i) \quad \text{to} \quad (A + r_i), \quad i = 1, 2, 3, \ldots$$

in [0,1]   in $(-\infty, \infty)$.

Note that **[62]** tells us there is a one-to-one correspondence between the ith piece $A + q_i$ in [0,1], and its new position, as indicated by the ith piece $A + r_i$ in $(-\infty, \infty)$. (See Figure 3, item **[63]** following.)

But, the reordered pieces, $A + r_i$, now cover *all* of the real line! Let us see why the countably many sets, $A + r_i$, form a partitioning of *all* of $(-\infty, \infty)$.

[63]

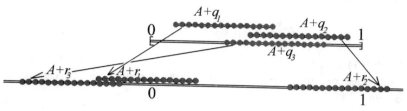

FIGURE 3
Set $A + q_i$ is moved to $A + r_i$, $\qquad i = 1, 2, 3, \ldots$.

Step 7. **Check that the countably many, (reordered) pieces, $A + r_i$, cover the whole real line.**

[64] Weak Form of Banach-Tarski Paradox. The countably many partitions,

$$A + q_i : i = 1, 2, 3, \ldots,$$

that cover a subset of [0,1], when reordered by moving

$$A + q_i \quad \text{to} \quad A + r_i,$$

produce a countable set of (shifted) partitions,

$$A + r_i : i = 1, 2, 3, \ldots,$$

that cover all of $(-\infty, \infty)$.

There are no gaps. We will show that the countably many pieces, $A + r_i$, cover all of $(-\infty, \infty)$. That is, any x whatever will belong to one of the reordered pieces, $A + r_i$ in $(-\infty, \infty)$. Accordingly, let's shift x by a rational distance, p, so it falls in the interval [0, 1/2]. That is,

$$(x + p) \in [0, 1/2].$$

Now every number in [0, 1/2] is in *some* equivalence set B_α as shown in Figure 2, item **[59]**. Therefore, $(x + p)$ is a rational jump or distance, p' say, to its representative equivalent a, which the Axiom of Choice selected to produce core set A in $[0, 1/2]$. In other words, for some rational p',

$$(x + p) + p' = a \in A.$$

From **[61]**, every rational number, such as $-(p + p')$ is of the form r_i for some integer i. Add $-(p + p') = r_i$ to both sides of the equation above, to obtain

$$x = a - (p + p')$$

$$= a + r_i \in A + r_i.$$

This shows that any x belongs to some reordered piece $A + r_i$ in the countable partition of $(-\infty, \infty)$. Our randomly chosen real number x has not escaped being covered by some shifted core set $A + r_i$.

At this point, Anvil Willie looked a little stunned as he stared in Hutch's direction. Cordelia, looking at the far wall, was nodding in silence. I just looked down at my coffee.

"I thought you would be impressed, not depressed," Hutch proclaimed. "Oh no, we are not depressed," Anvil Willie responded. "We are very much impressed to learn that reordering pieces of a little set can produce a huge set. It's just that I was much more comfortable before I met you—when I could believe that mathematics was just adding up numbers."

—John de Pillis
from *Starlight Café Conversations: An Illustrated Dictionary
from Table Seven.* (See pg. 319.)

[65] Drifting Toward Urban Legend

*A **Very Good Story:** It is often stated that Alfred Tarski considered the Axiom of Choice to be rubbish. In fact, it is said, he often trotted out the Banach-Tarski paradox to prove the axiom's disastrous and absurd consequences. Mathematicians, instead of being shocked by such a "ridiculous" and non-intuitive result, took perverse glee in embracing it all the more.*

It is a good story. For first-hand facts, however, read on.

—JdP

[66] I was a graduate student at Stanford in 1951–55, [while] my husband was a faculty member there and [neither of us recalls] Tarski ever making any derogatory remark about the Axiom of Choice. The Banach-Tarski paradox was published in 1924. Since then, Tarski himself wrote quite a few papers about the Axiom of Choice. For example, there are references to these papers in *Equivalents of the Axiom of Choice* by H. Rubin and J.E. Rubin, North Holland, 1985 and *Consequences of the Axiom of Choice* by P. Howard and J.E. Rubin, AMS, 1998.

—Jean E. Rubin*, Purdue University
(private communication)

[67]

Avoiding the Axiom of Choice: After all the sound and fury about unreasonable consequences of the Axiom of Choice, the Banach-Tarski paradox has now been proved without it. This was done by Randall Dougherty and Matthew Foreman, in "Banach-Tarski decompositions using sets with the propery of Baire," J. Amer. Math. Soc. vol. 7,1, (1994), pp. 75–125.

—JdP

AXIOMS AND DEFINITION

[68] Critical Thinking

Whenever someone tries to persuade you to adopt a certain point of view, then a fair question is, "what led you to that conclusion?" If language matters at all, then a clear statement of assumptions, axioms, or hypotheses is an essential characteristic of "critical thinking." The following illustrates what can happen when we casually accept conclusions without isolating or understanding the basic assumptions.

—JdP

Note: See also "Anatomy of Thought," items **[20]**–**[35]**.

The scouts asked their scoutmaster in autumn if the winter was going to be cold or not. Not really knowing the answer, the scoutmaster replied that the winter was going to be cold and that the members of the troop were to collect wood to be prepared.

Being a good leader, he then went to a nearby phone booth and called the National Weather Service and asked, "Is this winter to be cold?"

The man on the phone responded, "This winter will be quite cold indeed."

So the scoutmaster went back to speed up his scouts to collect even more wood to be prepared. A week later he called the National Weather Service again, "Is it going to be a very cold winter?"

"Yes," the man replied, "it's going to be a very cold winter."

So the scoutmaster went back to his scouts and ordered them to gather every scrap of wood they could find. Two weeks later he called the National Weather Service again: "Are you absolutely sure, that the winter is going to be very cold?"

"Absolutely," the man replied, "the scouts are collecting wood like crazy!"

—Anonymous

[69]

In a paper bearing the title,
"Axioms are positively vital,"
    A logician unnamed,
    Was the one who proclaimed
"Counting's a Peano recital."

—Paul Ritger

[70] Questions that pertain to the foundations of mathematics, although treated by many in recent times, still lack a satisfactory solution. The difficulty has its main source in the ambiguity of language.

—Giuseppe Peano (1858–1932)
Opening of the paper "Arithmetices Principia"
in which he introduced axioms for the integers.

(For more on ambiguity, see items **[99]**–**[102]**.)

[71] The axioms of geometry. . . are only definitions in disguise. What then are we to think of the question: Is Euclidean geometry true? It has no meaning. We might as well ask if the metric system is true and if the old weights and measures are false; if Cartesian coordinates are true and polar coordinates are false. One geometry cannot be more true than another; it can only be more convenient.

—Henri Poincaré (1854–1912)
Quoted in Marvin J. Greenberg's *Euclidean and non-Euclidean Geometries: Development and History*,
3rd ed. W. H. Freeman & Co., Oct. 1985.

Note: See also J. L. Heilbron, item **[594]**, discussing the centuries required to develop the deductive process.

Shermer's Ten Rules for Baloney Detection

[72]

Here are Michael Shermer's ten tips for what he calls "baloney detection." These rules indicate the boundaries between science and pseudoscience.

—*JdP*

1. How reliable is the source of the claim?
2. Does this source often make similar claims?
3. Have the claims been verified by another source?
4. How does the claim fit with what we know about how the world works?
5. Has anyone gone out of the way to disprove the claim, or has only supportive evidence been sought?

6. Does the preponderance of evidence point to the claimant's conclusion or to a different one? The theory of evolution, for example, is proved through a convergence of evidence from a number of independent lines of inquiry. No one fossil, no one piece of biological or paleontological evidence has "evolution" written on it; instead tens of thousands of evidentiary bits add up to a story of the evolution of life.

7. Is the claimant employing the accepted rules of reason and tools of research, or have these been abandoned in favor of others that lead to the desired conclusion?

8. Is the claimant providing an explanation for the observed phenomena or merely denying the existing explanation?

9. If the claimant proffers a new explanation, does it account for as many phenomena as the old explanation did?

10. Do the claimant's personal beliefs and biases drive the conclusions, or vice versa?

—Michael Shermer,
founding publisher of *Skeptic* magazine (www.skeptic.com)
Scientific American, Nov 2001.

Definitions for the Scientist

[73]   Mathematics, once fairly established on the foundation of a few axioms and definitions, as upon a rock, has grown from age to age, so as to become the most solid fabric that human reason can boast.

—Thomas Reid (1710–1796)

[74]   The errors of definitions multiply themselves according as the reckoning proceeds; and lead men into absurdities, which at last they see but cannot avoid, without reckoning anew from the beginning.

—Thomas Hobbes, (1588–1679)
In James R. Newman (ed.) *The World of Mathematics*,
New York: Simon and Schuster, 1956.

[75]   To be sure, mathematics can be extended to any branch of knowledge, including economics, provided the concepts are so clearly defined as to permit accurate symbolic representation. That is only another way of say-

ing that in some branches of discourse it is desirable to know what you are talking about.

—James R. Newman (ed)
The World of Mathematics, New York: Simon and Schuster, 1956

[76] It hath been an old remark, that Geometry is an excellent Logic. And it must be owned that when the definitions are clear; when the postulate cannot be refused, nor the axioms denied; when from the distinct contemplation and comparison of figures, their properties are derived, by a perpetual well-connected chain of consequences, the objects being still kept in view, and the attention ever fixed upon them; there is acquired a habit of reasoning, close and exact and methodical; which habit strengthens and sharpens the mind, and being transferred to other subjects is of general use in the inquiry after truth.

—George Berkeley (1685–1753)
The Analyst: a Discourse addressed to an Infidel Mathematician,
essay 2, on geometry and improvement of the mind.

Definitions for the Non-Scientist

[77] "When I use a word," Humpty Dumpty said, in a rather scornful tone, "it means just what I choose it to mean—neither more nor less."

"The question is," said Alice, "whether you can make words mean so many different things."

—Lewis Carroll (1832–1898) pseud. of Charles Lutwidge Dodgson,
Through the Looking Glass.

[78] Gore Vidal and Other Humpty Dumpty Thinkers: Reality Trumps Satire.

Language, the conduit of ideas, matters. It has been said that clear definitions are all very well for the physicist, the mathematician, the astronomer, or the chemist. Others maintain, however, that in natural language, the role of definitions is different. An over-emphasis on precision may rob language of its beauty and poetry.

Perhaps there is a larger class with an obligation to be clear in the meaning of their words—the "Persuaders"—those who strive to have others accept their point of view. (The Persuaders include scientists.) The following is an excerpt of a discussion of this very topic by the occupants of Table Seven of the Starlight Café. (See pg. 319.)

—JdP

"Humpty Dumpty thinkers are really out there, Hutch," Anvil Willie exclaimed as he looked down at the newspaper. "In this article, the word 'justice' isn't used in any sense I'm familiar with. It must mean something else—something only the author wants it to mean." (See item **[77]** for Humpty Dumpty's actual comment.)

[79] "Humpty Dumpty? Willie, what are you talking about?" Cordelia asked. Anvil Willie read out loud, quoting Gore Vidal (item **[84]**). His comment on mass murderer, Timothy McVeigh was:

"The boy [Timothy McVeigh] has a sense of justice.... That's what attracted me to him."

—Gore Vidal, quoted in an article by Martin Kettle,
National Guardian, Monday May 7, 2001. See also guardian.co.uk
and KOCO-TV channel 5, channeloklahoma.com 5 May 2001.

After a short pause, Hutch responded, "Well, my mathematics and science classes show me that clear definitions are very important. I don't know about you all, but this article convinces me that non-mathematicans also would benefit from clear definitions."

[80] "But wait," Cordelia, objected, "In natural language, you can't pin down exact meanings of every word. Human communication is not like science, you know."

(To support her point, Cordelia quoted from a poem, "Skunk Cabbage," which included the line,

> Digging a grave at the edge of the bog,
> the muck in love with my shovel, sucking it in,
> down I go.

—Rennie McQuilkin
from *We All Fall Down*, Swallow's Tale Press, ©1987.)

Would this poem be enhanced or diminished if "muck," "in love" and "shovel" were clearly defined? Or is it the creative *combination* of well-defined words that gives a poem its quality? Hutch recalled a comment from physicist Paul Dirac on this point (see item [399]) as well as Newton's unfriendly comment about poetry (see item [401]). Interesting as this was, we agreed to discuss poetry at a later time at Table Seven.

Resuming our conversation on "justice," I took out my pocket dictionary and provided the following working definition:

[81] Definition: **Justice** is the enactment or assignment of appropriate rewards or punishments resulting from certain action or actions."

Hutch thought for a moment. Then she said, "You know, there are really only two possibilities. Either

(a) Vidal is using 'justice' in its usual sense, or,

(b) Vidal is using 'justice' as Humpty Dumpty would, with a meaning known only to Vidal himself."

"And what do you conclude from these two possibilities?" Anvil Willie asked.

Hutch continued, "Well, in the first case, when 'justice' is used in the usual sense, Vidal is arguing that McVeigh's murder of 168 innocent people is an appropriate reward or punishment for a Government action that McVeigh disapproves of.

At this point, Cordelia winced, clearly upset by an argument that describes the murder of innocent people as "appropriate."

"In the remaining case, when 'justice' is used with a new meaning," Hutch continued, "Vidal is using 'justice' to mean whatever he wants it to—just like Humpty Dumpty." (See Humpty Dumpty, item [77].)

"Yeah. Satire, take a back seat," Anvil Willie said.

We all sat there not knowing what he meant. When we asked for a clarification, Anvil Willie replied, "I thought Lewis Carroll's Humpty Dumpty satire, like all satire, was supposed to be an exaggeration. But it ain't! Humpty Dumpty Persuaders are *among* us. I repeat,

[82]   *Satire, take a back seat, here comes Reality.*

"Oh, let's be serious," Cordelia protested. "Since Vidal is not a scientist, then what harm is being done if, occasionaly, he is careless with definitions? He's just talking—he's not building an atom bomb!"

"That's not exactly true," Anvil Willie replied. "Language matters. When you distort ideas, you are, figuratively speaking, building an atom bomb. Vidal is a Persuader, after all. Having said that, I think we should agree on a definition of 'persuader.' "

[83] Definition: A Persuader is one who uses words with the intent of changing opinions, beliefs, and values.

I decided to add my voice to the discussion. "Well, if a Persuader is going to enter my head with words in order to rearrange the furniture of my mind, then I sure want to know what those words mean!"

Cordelia surprised us all when she said, thoughtfully, "My values and beliefs are very private and precious to me. When someone wants the privilege of changing those areas, then they have the obligation to let me know—in the clearest terms —what their message is. Yes, language does matter."

To illustrate that using words to change peoples' core values can have powerful consequences—for good and for ill—we raised the question:

What if Adolph Hitler had his feet held to the fire? What if he was required to define 'Aryan superiority.' We would know whether 'superiority' meant better hospitals and schools. Or whether Hitler was talking about domination—superiority in the bullying of others. It would have been useful to have Hitler, the Persuader, define his terms.

As Hutch took a sip of her coffee, she peered at Anvil Willie over the rim of her cup, saying, "Interesting newspaper article, Willie. Even more interesting than the sports page, I bet."

—John dePillis from *Starlight Café Conversations: An Illustrated Dictionary from Table Seven* (see pg. 319.)

[84]

*Writer Gore Vidal (b. circa 1925, West Point, N.Y.),
grandson of Senator Thomas Gore of Oklahoma, has been
called "an acerbic observer of the contemporary
American scene" and "an acute commentator on the*

nation's history." Vidal defends his quote of item **[79]** *in his book,* Perpetual War for Perpetual Peace, *Thunder's Mouth Press, 2002.*

—JdP

[85]

Dr. John E. Mack, *Harvard Research Psychiatrist, a Pulitzer Prize winner for the 1977 biography of T. E. Lawrence, more recently, has engaged in research attempting to support claims of those who say they have been abducted by aliens. However, Dr. Mack provides no definitions and he dismisses the need for evidence. (See the following paragraph.)*

—JdP

[Dr. John E. Macks'] theory: Some powerful intelligence is trying to intervene in human affairs.... Mack said he considers critics' demands for more physical evidence a distraction from his main inquiry into why human beings are being contacted and by whom.... Mack brushes off...skeptics, saying his job is to persuade society that [people claiming to have been abducted by space aliens] deserve respect. He gives them the appellation, "authentic witnesses" and says they carry a message that Earthlings would ignore only at our peril.

—Michael Lucas, *Los Angeles Times*, Sept. 4, 2001, pg. E2, in an article on Dr. John E. Mack who supports claims of those who say they have been abducted by aliens.

[86]

Kuhn's influence on the non-scientist: *A non-scientist who wants to learn about the history and development of scientific thought is likely to consult Thomas Kuhn's book,* The Structure of Scientific Revolutions, *University of Chicago Press, 3rd edition, 1996 (hereafter abbreviated as SSR). The importance of SSR to non-scientists can be measured by the number of citings in the Arts and Humanities Citation Index, which lists all citations in the major humanities journals. Recently, SSR was the most*

often-cited single work (although Kuhn was not the most cited author), far outnumbering citings of Northrop Frye's Anatomy of Criticism *and James Joyce's* Ulysses.

The continuing influence of Kuhn, the Persuader, is confirmed by Daniel P. Maloney, who writes:

> *Hailed by the poststructuralist left, wielded by feminists and fundamentalists alike, and hated by most practitioners of the field it purports to explain, Thomas Kuhn's* The Structure of Scientific Revolutions *is perhaps the single most important work on the nature of rationality since Descartes'* Meditations.
>
> —*Daniel P. Maloney, associate editor.*
> First Things, *no. 101, March 2000, pp. 53–55.*

> —*JdP*

[87] Thomas Kuhn and the Value of Definition

Kuhn's use of definition: *If Kuhn has become science's spokesman in the eyes of the non-scientist (i.e., a Persuader), then it is appropriate to ask what importance he places on clear and unambiguous definitions. Does language matter? To see the degree of attention Kuhn pays to unambiguous definitions, consider the following passage in his own words:*

> —*JdP*

Notes. See also J. L. Heilbron, item **[594]**, who notes that centuries were required to produce the deductive process; E. T. Bell, item **[409]**, on the need for clarity of assumption; and James Franklin, *The New Criterion* Vol. 18, No. 10, June 2000, www.newcriterion.com, for Kuhn's influence on non-scientists.

See also items **[581]**–**[583]** for a discussion of how well Kuhn represents the practicing scientist's view of progress.

For balance, see also item **[99]** for the fortunate misunderstanding by Alexander Graham Bell, and items **[100]**–**[102]** for cases of unavoidable ambiguity.

[88]   Men argue that psychology, for example, is a science because it possesses such and such characteristics. Others counter that those characteristics are either unnecessary or not sufficient to make a field a science. Often
great energy is invested, great passion aroused, and the outsider is at a loss
to know why. Can very much depend upon a definition of 'science?' Can
a definition tell a man whether he is a scientist or not? If so, why do not
natural scientists or artists worry about the definition of the term?

—Thomas Kuhn (1922–1996)
The Structure of Scientific Revolutions, University of Chicago Press,
pg. 160, 3rd edition, 1996.

[89] Kuhn's 22 Definitions of "Paradigm"

*I say "paradigm," you say "model:" In SSR, Kuhn
introduces the term "paradigm" to frame his core ideas.
(Is "paradigm" better or clearer than "model"?) The
word "paradigm" has gained such currency and
popularity that William Safire's New Political Dictionary
has an article on "paradigm shift." As another example,
note that Persuaders (in the popular media, for example)
will often say that new evidence produces not a "new
model," but a "paradigm shift."*

*What importance does Kuhn, the Persuader, place on
providing the reader with a clear definition of his key term
"paradigm?" Once again, we may ask, does language
matter? Here is Kuhn's response in his own words:*

—JdP

[90]   [Let us] turn to paradigms and ask what they could possibly be. My
original text leaves no more obscure or important question. One sympa-
thetic reader, who shares my conviction that 'paradigm' names the central
philosophical elements of the book, prepared a partial analytic index and
concluded that the term is used *in at least twenty-two different ways....*
[After some editorial work, however, only] *two very different usages* of the
term would remain. (*Italic emphasis: JdP*)

—Thomas Kuhn (1922–1996)
Structure of Scientific Revolutions, University of Chicago Press,
pp. 181–182, 3rd edition, 1996.

[91] Alan Sokal's Puckish Hoax

Testing the standards: Who would have guessed that clarity of terminology would not only be ignored, but would be actively opposed by scholars who publish a research journal? Here is the story of NYU physicist, Alan Sokal, who wrote two famous articles.

- *The deliberately phony one [Transgressing the Boundaries: Toward a Transformative Hermeneutics of Quantum Gravity, Social Text, 1996 spring/summer issue]. This post-modernist academic journal maintains that truths are not objective, but are related to and validated by culture. (For example, the speed of light in China is different from that in the U.S.) Sokal became a "valid" culture of one and wrote a paper filled with Sokal-speak. It was accepted as scholarly and valid.*

- *An honest, albeit gleeful, confession of his prank [Lingua Franca, May 1996].*

—JdP

[92]   Is it now dogma in cultural studies that there exists no external world? Or that there exists an external world but science obtains no knowledge of it? In the second paragraph I declare, without the slightest evidence or argument, that "physical 'reality' [note the scare quotes]...is at bottom a social and linguistic construct...." Anyone who believes that the laws of physics are mere social conventions is invited to try transgressing those conventions from the windows of my apartment. (I live on the twenty-first floor.)...

What concerns me is the proliferation, not just of nonsense and sloppy thinking *per se*, but of a particular kind of nonsense and sloppy thinking: one that denies the existence of objective realities...

There is a real world; its properties are not merely social constructions; facts and evidence do matter. What sane person would contend otherwise?

—Alan Sokal, physicist, NYU
Lingua Franca May/June 1996.

[93] Importance of Alan Sokal's Concerns

Too many academics [who are] secure in their ivory towers and insulated from the real-world . . . seem blind to the fact that non-rationality has historically been among the most powerful weapons in the ideological arsenals of oppressors.

—David Whiteis

Note: See also Gore Vidal and Other Humpty Dumpty Thinkers, item **[78]**.

Three Rebuttals to Alan Sokal's Hoax

*You may wish to measure the following three rebuttals against the characteristics of faulty argument (mind-reading, motive analysis, personal attack) described in items **[2]** and **[3]**.*

—*JdP*

[94]   What Sokal's hoax reveals is his anxiety as a scientist about this empire of knowledge, as well as a rather ignorant and chortlingly condescending relation to complicated philosophical, sociological, and theoretical positions.

—George Levine, Director, Center for the Critical Analysis of Contemporary Culture, Rutgers University, *Lingua Franca*, July/August 1996.

[95] [It] is Alan Sokal, not his targets, who threatens to undermine the intellectual standards he vows to protect. No scientist begins his task by inventing anew the facts he will assume. They are all given by the tradition of inquiry he has joined, and for the most part he must take them on faith. And he must take on faith, too, the reports of his colleagues. [Sokal] carefully packaged his deception so as not to be detected except by someone who began with a deep and corrosive attitude of suspicion.

—Stanley Fish, professor of English, University of Illinois, Chicago, formerly executive director of the Duke University Press, which publishes *Social Text. New York Times*, May 21, 1996.

[96] Our main concern is that readers. . .are not persuaded by the Sokal stunt that this is simply an academic turf war between scientists and humanists/social scientists, with each side trying to outsmart the other.

—Bruce Robbins and Andrew Ross, co-editors, for *Social Text Lingua Franca*, July/August 1996.

Note: For other accounts of Sokal's Hoax see *Did Adam and Eve Have Navels?* by Martin Gardner, W. W. Norton & Co., 2000, ch. 14, and *Fashionable Nonsense* by Alan Sokal and Jean Briemont, Picador, USA, 1999.

THE ARROGANCE OF SCIENTISTS

[97]

When non-scientists (mis)perceive science. Huston Smith, a religious scholar and author of Why Religion Matters, *has been accused of arguing that scientists are arrogant in their claim that their scientific truth is the only truth. But when Smith says, "Science didn't really say this," (see full quote in item* **[98]** *following) then Smith's complaint seems directed not at scientists at all, but at the faulty mind-reading abilities of non-scientists who imagine or perceive arrogance of scientists. (See item* **[2]** *on the role of mind-reading in bad conversation.) So much for reliability when A (a Persuader) tells B (you) what C (the scientist) is thinking.*

Words of actual scientists. In item [210], Stephen J. Gould comments on the separate roles of science and religion, James Newman, in item [75], and George Berkeley in item [76] note the benefits of clarity in all areas of discussion. For acknowledgement of the built-in uncertainty in science, see remarks of K.C. Cole in item [122]. Richard Feynman, in item [126], gives his views of uncertainty in science and life, and, of course, the provable statement by Kurt Gödel in item [324] that, even in mathematics, we are humbled by the fact that there always are true statements we shall never prove.

This is arrogance?

—JdP

[98]   Science has become the revelation of our time. And ... it should be with regard to the material world.... The slip is that we have turned science into scientism—scientism being defined as the assumption that science is the only reliable way of getting at truth, and that only the kinds of things it tells us about really exist.

...Science didn't really say this, but because its power derived from attending to the material aspects of nature, and because that power is great and effective and gave us many benefits, the outlook of modernity is un-precedentedly materialistic.

—Huston Smith from an interview with Michael Toms in the
publication, *What is Enlightenment?*
Lenox, MA, Issue 11 (*Can Science Enlighten Us?*)

Note: See also items [249]–[251] in the section on humility.

NOW FOR SOME IRONY: BELL'S VALUABLE BLUNDER

[99]

For all our emphasis (if not self-righteousness) invested in clarity (see items [68]–[98]), wouldn't you know, there exists at least one case in which misunderstanding of language proved to be beneficial!

According to the PBS history series, "The American Experience," Alexander Graham Bell, while studying at the University of London, became intrigued by a thesis of the German physicist Hermann von Helmholtz entitled, On The Sensations of Tone. *In this work, von Helmholtz asserted that vowel sounds could be produced by a combination of electrical tuning forks and resonators.*

Bell couldn't read German. No matter. Such as he could, he hungrily consumed von Helmoltz's writings. It did, however, lead to what he would later describe as a "very valuable blunder."

Somehow, Bell (mis)read von Helmholtz's findings to mean that vowel sounds could be transmitted over a wire. He would later say of this misunderstanding, "It gave me confidence. If I had been able to read German, I might never have begun my experiments in electricity."

—JdP

(For more of Von Helmholz's thoughts, see item **[379]** concerning music and intelligence.)

THREE EXAMPLES OF UNAVOIDABLE AMBIGUITY

Clarity in meaning, *however desirable, is not always easy or even possible to achieve in natural language. Here are some examples.*

[100]   The following is a sentence, with two entirely different meanings. It has come to be a classic in the field of logic and artificial intelligence.

Time flies like an arrow.

Meaning 1. *Time (a verb) flies (a noun) in flight just as you time arrows in flight.*

Meaning 2. *Time (the noun) passes as quickly as an arrow does.*

[101]   *The following was found in a jewelry shop window. Its two meanings are opposites of each other.*

Nothing makes you feel as good as gold.

Meaning 1. *If you are looking for something to make you feel really good (as good as gold, say), then forget it—nothing will do the job.*

Meaning 2. *If you are looking for something to make you feel good, then gold will do the job—and nothing does it better. (Presumably, this is the jeweler's intended meaning.)*

[102]   *In writing a letter of recommendation for your recently released employee, Hector, you might use the ambiguous sentence,*

> You will be lucky if you can get
> Hector to work for you.

Examples like [100]–[102] *abound in natural language. They show us why context is important, why word-for-word translation is difficult, and why accurate machine translation of one language into another is bound to be a daunting task.*

Lest we think that ambiguity is exclusive to natural language and that mathematics suffers no such flaw, read item [70] *for Peano's comment and item* [227] *for Bertrand Russell's remarks on ambiguity and circularity of definitions.*

—JdP

ARISTOTLE AND AN AVOIDABLE AMBIGUITY

Careless Definitions Can Deceive. *Here is a quote you will find in the literature that purports to state Aristotle's definition of democracy.*

[103]   Democracy is when the indigent, and not the
men of property, are the rulers.

These words suggest that Aristotle held democracy in some contempt.

But it's all a matter of definition. Aristotle's notion of democracy is far different from the high-minded notion of consensual democracy to which we adhere.

Consulting the source (item [104]), we see that Aristotle's definitions and categories of governments have everything to do with value judgments.

First, there are "good" rulers who regard the common interest. These rulers correspond to the three categories:

(a) monarchies (one person rules),

(b) aristocracy (a few people rule),

(c) polity (citizens at large rule).

Secondly, there are "bad" rulers who pervert the categories above. The three perverted categories are:

(a) tyranny, a perversion of monarchy,

(b) oligarchy, a perversion of aristocracy,

(c) democracy, a perversion of polity.

Here, in Aristotle's own (translated) words, are the details.

—JdP

[104]   Of forms of government in which one rules, we call that which regards the common interests, monarchy; that in which more than one, but not many, rule, aristocracy (and it is so called, either because the rulers are the best men, or because they have at heart the best interests of the state and of the citizens). But when the citizens at large administer the state for the common interest, the government is called a polity. And there is a reason for this use of language.

Of the above-mentioned forms, the perversions are as follows: of monarchy, tyranny; of aristocracy, oligarchy; of polity, democracy. For tyranny is a kind of monarchy which has in view the interest of the monarch only; oligarchy has in view the interest of the wealthy; democracy, of the needy: none of them the common good of all. Tyranny, as I was saying, is monarchy exercising the rule of a master over the political society; oligarchy is when men of property have the government in their hands; democracy, the opposite, when the indigent, and not the men of property, are the rulers. (Compare this last sentence with item [103].)

—Aristotle (384–322 B.C.) from *The Politics*,
Book III, circa 340 B.C.

Note: See items [769] and [770] for George Washington's role in securing a consensual democracy.

RADIUS OF REALITY: HUMPTY DUMPTY AND BEYOND

[105]

What's going on here? Gore Vidal, in Humpty Dumpty fashion, renders the term "justice" as meaningless (item [79]) while Dr. John Mack considers demands for more physical evidence to be a distraction (item [85]). Thomas Kuhn, who educates non-scientists on how scientists think (in spite of scientists' objections), seems unaware that clear definitions are of any importance at all (items [88]–[90]). Since all points of view are equally valid (who are we to judge?), physicist Alan Sokal is able to publish an intentionally meaningless article in the post-modernist journal, Social Text *(items [91]–[96]). Finally, if Aristotle's statement [103] is better understood only after terms are properly established (item [104]), then don't definitions matter? Why is clarity not universally endorsed?*

Surprisingly, clarity is a casualty of distance. The fundamental question is; how far can we cast our net of understanding into the sea of reality?

If the answer is not far, if our radius of knowable reality has zero length, then there can be no "out there" where standards exist for everybody. Moreover (this self-centered argument continues), listeners will only (re)interpret remarks of others in terms of their personal language. So why waste efforts on trying to be clear?

***Ludwig Wittgenstein** (1889–1951) (who modified his opinions often) was one of the earliest philosophers who started us thinking that there is no knowability or validity beyond language itself. Working only with language, we are casting a zero-radius net that captures a "picture of*

facts" (the way a photo produces a representation of
"reality"), but that's it. There are, Wittgenstein asserts, no
other legitimate uses of language such as evaluating
ethical or moral concepts like altruism, evil, or murder.
Such judgments are individual in-the-head constructs that
are valid (maybe) for one but not for all. Therefore, value
judgments or standards are nonsense, irrelevant to a
group or society at large.

In contrast, most scientists hold that we can and should
explore beyond ourselves in an unbounded sea of reality
(See item **[687]** for Henry James' comment on absolute
truth.) This assertion allows for a richer set of realities,
namely those that are physical, intellectual, emotional and
spiritual. (Items **[386]**–**[392]**, **[417]** speak to the
effectiveness of mathematics in understanding nature.)

The following quote reflects Wittgenstein's belief that we
live in a private "virtual" reality, each of which is created
through our own language games. He denies any
possibility of an absolute truth—including whatever truth
may lie even in his own writings as we see in item **[106]**!

—JdP

Note: See item **[324]** for the Incompleteness Theorem of Kurt Gödel; that says:

> There are true statements "out there" in mathematics that are unknowable in the sense that they forever lie beyond the reach of all mathematical proofs.

[106] My propositions serve as elucidations in the following way: anyone who understands me eventually recognizes them as nonsensical, when he has used them—as steps—to climb up beyond them. (He must, so to speak, throw away the ladder after he has climbed up it.) He must transcend these propositions, and then will he see the world aright. What we cannot speak about we must pass over in silence.

—Ludwig Wittgenstein (1889–1951)
Tractatus Logico-Philosophicus (1921) 6.54, 7

Note: For the emotional side of Wittgenstein, see item **[173]**.

B

THE BRAIN

[107]   If the human brain were so simple that we could understand it, we would be so simple we couldn't.

—Emerson M. Pugh (as quoted by George E. Pugh, Emerson's son in *G.E. Pugh, The Biological Origin of Human Values*, p. 154, 1977.)

[108]   Keeping an open mind is a virtue—but as the space engineer James Oberg once said, not so open that your brains fall out.

—Carl Sagan (1934–1996), *The Demon-Haunted World.*

*"You're telling me the Cubs will
win the pennant this year?
And you call yourself a brain?!?"*

[109] In proportion to our body mass, our brain is three times as large as that of our nearest relatives. This huge organ is dangerous and painful to give birth to, expensive to build and, in a resting human, uses about 20 percent of the body's energy even though it is just 2 percent of the body's weight. There must be some reason for all this evolutionary expense.

—Susan Blakemore (from "Me, Myself, I,"
New Scientist, March 13, 1999.)

[110] I, George Bush, President of the United States of America, do hereby proclaim the decade beginning January 1, 1990, as the Decade of the Brain. I call upon all public officials and the people of the United States to observe that decade with appropriate programs, ceremonies, and activities.

—George H. W. Bush (U.S. President 1988–1992.)
(from Presidential Proclamation 6158, July 17, 1990.)

[111] My hand moves because certain forces—electric, magnetic, or whatever 'nerve-force' may prove to be—are impressed on it by my brain. This nerve-force, stored in the brain, would probably be traceable, if Science were complete,* to chemical forces supplied to the brain by the blood, and ultimately derived from the food I eat and the air I breathe.

—Lewis Carroll (1832–1898) pseud. of Charles Lutwidge Dodgson
(from *Sylvie and Bruno*, 1890.)

*Note: *Although Lewis Carroll discusses the possible incompleteness of science, logic itself, a basic tool in our understanding of science, is intrinsically incomplete! See Kurt Gödel's 1931 Incompleteness Theorem, item* [324].

—JdP

BORE

[112] The secret of being a bore is to tell everything.

—Voltaire (1694–1778)
Sept Discours en Vers sur l'Homme, 1738.

Note: See also Evariste Galois, item [154], on expositors revealing their areas of ignorance.

C

CALCULUS

Note: See also nonintuitive calculus examples, items [293]–[296].

[113] If a nonnegative quantity was so small that it is smaller than any given one, then it certainly could not be anything but zero. To those who ask what the infinitely small quantity in mathematics is, we answer that it is actually zero. Hence there are not so many mysteries hidden in this concept as there are usually believed to be. These supposed mysteries have rendered the calculus of the infinitely small quite suspect to many people. Those doubts that remain we shall thoroughly remove in the following pages, where we shall explain this calculus.

—Leonhard Euler (1707–1783) from
A Source Book in Mathematics 1200–1800,
D. J. Struik, editor, Princeton University Press, 1986.

[114] The Derivative Song.　Tune: *There'll be Some Changes Made*

> You take a function of x and you call it y,
> Take any x-nought that you care to try,
> You make a little change and call it delta x,
> The corresponding change in y is what you find nex',
> And then you take the quotient and now carefully
> Send delta x to zero, and I think you'll see

That what the limit gives us, if our work all checks,
Is what we call dy/dx,
It's just dy/dx.
—Tom Lehrer, *The American Mathematical Monthly*,
vol. 81, 1974, p. 490.

[115]

ME
I am not, as a general rule, averse to Work,
but I do, I fear, have Moments of Inertia.
I go around in circles with a Torque,
and in beyond my depth with Liquid Pressure.
While the bulk o'me is levity,
I have a Center of Gravity.
—Katherine O'Brien, *The American Mathematical Monthly*,
vol. 73, no. 7, 1966, p. 732.

[116]   The calculus is the greatest aid we have to the appreciation of physical truth in the broadest sense of the word.

—W. F. Osgood

[117]   And what are these fluxions? The velocities of evanescent increments. And what are these same evanescent increments? They are neither finite quantities nor quantities infinitely small, nor yet nothing. May we not call them the ghosts of departed quantities?

—George Berkeley (1685–1753) *The Analyst: a Discourse
addressed to an Infidel Mathematician,
essay 35, on the need to accept fluxions or infinitesimals*

[118] [Calculus is] the outcome of a dramatic intellectual struggle which has lasted for twenty-five hundred years.

—Richard Courant (1888–1972)

[119] The analytical geometry of Descartes and the calculus of Newton and Leibniz have expanded into the marvelous mathematical method— more daring than anything that the history of philosophy records of Lobatchevsky and Riemann, Gauss and Sylvester. Indeed, mathematics, the indispensable tool of the sciences, defying the senses to follow its splendid flights, is demonstrating today, as it never has been demonstrated before, the supremacy of pure reason.

—Nicholas Murray Butler (1862–1947), educator, called "Nicholas Miraculous Butler" by his good friend, Theodore Roosevelt.

[120]

The Old Oaken Calculus Problem
How dear to my heart are cylindrical wedges,
when fond recollection presents them once more,
and boxes from tin by upturning the edges,
and ships landing passengers where on the shore.
The ladder that slid in its slanting projection,
the beam in the corridor rounding the ell,
but rarest of all in that antique collection
the leaky old bucket that hung in the well

the leaky old bucket, the squeaky old bucket,
the leaky old bucket that hung in the well.
—Katherine O'Brien, *The American Mathematical Monthly*,
vol. 73, no. 8, 1996, p. 881.

CERTAINTY

Note: See also "Mathematics/Truth," items **[404]–[414]**.

[121]   As far as the laws of mathematics refer to reality, they are not certain; and as far as they are certain, they do not refer to reality.
—Albert Einstein (1879–1955) In James R. Newman (ed.)
The World of Mathematics, New York: Simon and Schuster, 1956.

[122]   One of the most common misconceptions about science (frequently fostered, I'm sorry to say, by science writers) is that scientific truth can't be trusted because it is continually being revised. *Au contraire.* It can be trusted precisely because it is continually being revised.
—K. C. Cole, *Los Angeles Times* science writer,
A Hole in the Universe, Harcourt, 2001, p. 85.

[123]   There is no greater mistake than to call arithmetic an exact science. There are permutations and aberrations discernible to minds entirely noble like mine to perceive. For instance, if you add a sum from the bottom up, and then from the top down, the result is always different.
—Mrs. La Touche, *Mathematical Gazette*, vol. 12.

[124]   I can doubt everything, except one thing, and that is the very fact that I doubt."

—René Descartes (1596–1650)

Note: For more from René Descartes, see also items **[731]**–**[733]**.

[125]   "But," you may say, "none of this shakes my belief that 2 and 2 are 4." You are quite right, except in marginal cases—and it is only in marginal cases that you are doubtful whether a certain animal is a dog or a certain length is less than a meter. Two must be two of something, and the proposition "2 and 2 are 4" is useless unless it can be applied.

Two dogs and two dogs are certainly four dogs, but cases arise in which you are doubtful whether two of them are dogs. "Well, at any rate, there are four animals," you may say. But there are microorganisms concerning which it is doubtful whether they are animals or plants. "Well, then living organisms," you say. But there are things of which it is doubtful whether they are living organisms or not. You will be driven into saying: "Two entities and two entities are four entities."

"O.K., Mr. Philosopher. If you don't like the term 'animals,' then bring back a Bertrand Russell entity for supper!"

When you have told me what you mean by "entity," we will resume the argument.

—Bertrand Russell (1872–1970)

[126] Why We Are Here. I can live with doubt and uncertainty and not knowing. I think it's much more interesting to live not knowing than to have answers which might be wrong. I have approximate answers and possible beliefs and different degrees of certainty about different things, but I'm not absolutely sure of anything and there are many things I don't know anything about, such as whether it means anything to ask why we're here, and what the question might mean. I might think about it a little bit and if I can't figure it out, then I go on to something else, but I don't have to know and answer, I don't feel frightened by not knowing things, by being lost in a mysterious universe without having any purpose, which is the way it really is so far as I can tell. It doesn't frighten me.

—Richard P. Feynman, (1918–1988) *The Pleasure of Finding Things Out: The Best Short Works of Richard P. Feynman,* Jeffrey Robbins (ed.), p. 25. Citing transcript of BBC television program Horizon in 1981. (For statements of faith from other scientists, see items **[204]**–**[216]**.)

[127] Segall's Law: It's possible to own too much. A man with one watch knows what time it is; a man with two watches is never sure.

—Lee Segall, creator of the original *Dr. I.Q.* tv/radio program

[128] If the suggestion that space-time is finite but unbounded is correct, then the Big Bang is rather like the North Pole of the earth. To ask what happens before the Big Bang is a bit like asking what happens on the surface of the earth one mile north of the North Pole. It's a meaningless question.

—Stephen W. Hawking (b. 1942) In an interview with Timothy Ferris, Pasadena, CA, 1985 (For more on the Big Bang, see item **[167]**.)

Heisenberg Uncertainty Principle

[129]

Known as the founder of quantum mechanics, Werner Karl Heisenberg (1901-1976) stated the uncertainy principle at the age of 26.

The uncertainty principle in quantum theory says that the more precisely the position is determined, the less precisely the momentum is known in this instant, and vice versa. That is, there is always an uncertainty relation, or a

*trade-off, between knowledge of the position and the
momentum (mass times velocity) of a subatomic particle,
such as an electron.*

—*JdP*

[130] [Q]uantum mechanics revealed that the universe is inherently uncertain. If you try to pin down a particle to measure its properties, it slips
out of your grasp. The very act of holding it down to measure it, destroys
the properties you set out to measure—just as surely as holding a snowflake
in your hand melts the ice crystal before you can study its geometry....

*"It's the Heisenberg Bandit, Chief. All I can tell you is he's
going at 2500 rpm but I can't tell you exactly where he is!"*

Imagine a spinning coin. Is it heads or tails? It's neither, and it's both.
Or think of a rapidly spinning fan blade. Where, exactly, is the blade? In
the quantum world, it is everywhere at once—at least until you measure
it. Once you stop the blade, you can say precisely where it is. But in the
process, you have effectively "destroyed" its velocity.

—K. C. Cole, *Los Angeles Times* science writer,
A Hole in the Universe, Harcourt, 2001, pp. 78–79.

COMMON SENSE

[131] Lottery: A tax on people who are bad at math.

—Bumper sticker.

[132] Why is it that we never read a headline that says something like,
"Psychic wins the lottery!"?

[133]   Do not imagine that mathematics is hard and crabbed, and repulsive to common sense. It is merely the etherealization of common sense.

—William Thomson (later Lord Kelvin) (1824–1907)
as quoted in S.P. Thompson, *Life of Lord Kelvin* (London 1943).

[134]   Good sense is, of all things among men, the most equally distributed; for every one thinks himself so abundantly provided with it, that those even who are the most difficult to satisfy in everything else, do not usually desire a larger measure of this quantity than they already possess.

—René Descartes (1596–1650)
Discours de la Méthode, 1637

[135]   Common sense is, as a matter of fact, nothing more than layers of preconceived notions stored in our memories and emotions for the most part before age eighteen.

—Albert Einstein(1879–1955) quoted in E.T. Bell,
Mathematics: Queen and Servant of Science
(MAA *Spectrum Series*), December 1987.

[136]   Mathematics is often erroneously referred to as the science of common sense. Actually, it may transcend common sense and go beyond either imagination or intuition. It has become a very strange and perhaps frightening subject from the ordinary point of view, but anyone who penetrates into it will find a veritable fairyland, a fairyland which is strange, but makes sense, if not common sense.

—Edward Kasner, and James R. Newman, *Mathematics
and the Imagination*, New York: Simon and Schuster, 1940.

COMPUTERS

Mathematics—the Hidden Engine

Note: See items **[146]** and **[147]** for some early computer history.

[137]

Dazzling effects in computer graphics are more the results of mathematical algorithms and less the product of the optical camera. Surface-generating algorithms, for example, often depend on classical variational methods

and application of energy minimization. Nonetheless, the
foundation of these effects—the mathematics that produce
them—is generally ignored or unappreciated.

—JdP

I write the algorithms that make the computer sing. I'm the Barry Manilow of mathematics.

—Stanley Osher, University of California, Los Angeles
As reported in the *Los Angeles Times*, July 14, 1998, p. 1.

Note: See also Tony Chan, item **[349]**.

Programming and the Machine

Note: See also item **[530]** for Ron Graham's related remarks on juggling.

[138]   You think you know when you learn, are more sure when you can write, even more when you can teach, but certain when you can program.

—Alan J. Perlis

[139]   In the computer industry, there are three kinds of lies: lies, damn lies, and benchmarks.

—quoted in "The Free On-line Dictionary of Computing"
(15 Feb 1998)

Note: See also Benjamin Disraeli, item **[696]** commenting on statistical lies.

[140]   The Mac has a one-button mouse, the PC has a two-button mouse, and Linux has a three button mouse. Is there a message in this medium? If I ever figure it out, I will let you know.

—Al Kelley*, University of California, Santa Cruz

[141]   Lacking documentation, there is always imagination. (Applies equally well to public speakers and politicians who find themselves with an embarrassingly weak argument.)

—John de Pillis

[142]   The arithmetical machine produces effects which approach nearer to thought than all the actions of animals. But it does nothing which would enable us to attribute will to it, as to the animals.

—Blaise Pascal (1623–1662)
Penseés #340

[143]   Testing can show the presence of bugs. Testing can never show the absence of bugs.

—Edsger W. Dijkstra (1930–2002), a principal developer
of the ALGOL language. *A Discipline of Programming.*
Prentice-Hall, Englewood Cliffs, NJ, 1976, p. 6.

Note: See also item **[171]**, in which Karl Popper claims the only test of science lies in falsifying tests. Also, David Finkelstein, item **[676]**, sees physics developing through a progression of violating the laws and then modifying them.

*"I SEE THAT YOU ARE OF A GOOD CHARACTER. AND FOR
FIVE EXTRA DOLLARS, I CAN GIVE YOU ITS ASCII CODE."*

[144]   Niklaus Wirth, the inventor of PASCAL, has lamented that, whereas Europeans pronounce his name correctly (Ni-klows Virt), Americans invariably mangle it into (Nick-les Worth). Which is to say that Europeans call him by name, but Americans call him by value.

> —Jerry Pournelle, from the article, "Computing at Chaos Manor:
> Come to the Faire," *BYTE* magazine, vol. 10, no. 7,
> July 1985, pp. 309–339.

[145] Extending Collective Memory

"It's the memory extending ability that makes today's computers so powerful and indispensable, not their calculating ability," Anvil Willie said.

"What nonsense," Cordelia interrupted, "Why misname them 'computers' then if they are mainly doing something other than computing?"

"Look," Anvil Willie responded. "Think of some of the greatest inventions down the centuries: writing, the alphabet, the printing press, photography, the phonograph. These are nothing less than memory extenders. And if you have a collective memory which you can tap into at any time, then you have a solid base for building a civilization."

"Is extension of memory all that important?" asked Cordelia. "Well, let's think it through," Anvil Willie replied. "If your house is on fire, what are the first nonliving things you would try to save and recover? I would bet that photos and documents would be the first things you would try to save."

Anvil Willie paused to sip his coffee. "Think of it. Every city and every family has a memorial of some kind that revives or refreshes something in the past. And the word, 'memorial' itself has the same root as 'memory.' "

"I see. Visual and written memory extenders—they do seem to be important," Cordelia said thoughtfully, almost to herself.

Anvil Willie continued. "And don't forget. Astronomy would not have advanced at all if it weren't for photographs that freeze time so that the night sky can be studied at leisure." "Yes, I see your point, Willie. And we also have libraries and such," Cordelia acknowledged. Then, in a sudden burst, Cordelia added, "Wait! The Internet. That's information flow. That's not a memory thing."

"Really? Ever hear the term 'search engine'?" Anvil Willie replied. "What else do you tap into with a search engine if not a vast, retrievable, memory base?"

Cordelia sipped her coffee. With an air of finality, she said, "All right, Willie, who *I* am, as an individual, is molded and created by my *personal* memories. You're right about that. But now I have to come to grips with this: Who *we* are, as a civilization, is molded and created by the memory of your darned computers."

—John de Pillis from *Starlight Café Conversations: An Illustrated Dictionary from Table Seven*. (See pg. 319.)

Inventions with Minds of Their Own

[146]

*The computer, like the telephone, is used in ways that were unanticipated by the original inventors. For example, the **telephone** was developed by Alexander Graham Bell in order to ease the burden of the hearing impaired. (Bell was the son of a deaf mother and was husband and teacher to Mabel Hubbard who was also deaf.) Little did Bell anticipate that along with voice, wires would also electronically transmit e-mail, moving images, data and other forms of information. (See item [99] to read about*

*Bell's "valuable blunder" in the development of the
telephone.) Similarly, the computer was originally
designed to be a super calculator—to crunch numbers to
find solutions to differential equations and to plot missile
trajectories. The power of the computer's memory, which
we now see is fundamental in the functioning of search
engines and data bases, was not at first appreciated. Read
on for comments made in the 1957 film, Desk Set.*

—JdP

Peg, Peg, calm down! No machine can do our job.... They can't build
a machine to do our job. There are too many cross-references in this place.
I'd match my memory against any machine's any day.

—Katharine Hepburn, in the role of reference librarian
Bunny Watson, responding to the introduction of an electronic
brain in the work place. (From the 1957 film, *Desk Set*.)

[147]

How far advanced was computer technology in 1957, *the
year the film, Desk Set was released? (See item* **[146]**
above.)

*In 1957, an IBM team, led by John Backus, designed the
first successful high-level programming language,
FORTRAN. This language was purpose-built to solve
engineering and scientific problems.*

*In 1958, the IBM 7090, was introduced. This was the very
first computer to use the transistor as a switching device.*

—JdP

Embellishing Collective Memory: Apollo 13, April 1970.

[148]

*It can be argued that we all share a deep, inner need to
capture and plug in to collective memory so as to produce
an over-arching "bird's eye view" of where we came from,
what we have acquired, and where we should go—*

*sometimes stated as "the search for who we are." But
sometimes, we willingly sacrifice accuracy for warm and
pleasant romanticism. Even in technology.*

*For a case in point, consider the phrase "Failure is not an
option," which is attributed to NASA flight director Gene
Kranz as he was orchestrating the successful return of the
ill-fated crew of Apollo 13 on their return after orbiting
the moon. This catchy phrase was made up by screen
writers for actor Ed Harris who was playing the part of
Gene Kranz in the 1995 film, Apollo 13. What Gene Kranz
actually did say was,*

> *"We've never lost an American in
> space—we sure as hell aren't going to lose one
> now. This crew is coming home!"*

*Kranz himself liked his own quote as much as the made-up
phrase that he used for the title of his autobiographical
book,* Failure Is Not an Option, *Simon & Schuster, 2000.*

*But Kranz is honest in his book (for all the good it does)
where he accurately presents his original quote. In fact,
since Kranz has consistently and openly acknowledged
that he never said, "failure is not an option" it cannot be
said that the true statement is being hidden. Yet,
something in us impels us to reject the truthful recording
of Kranz' original, authentic, words in favor of the more
artfully crafted quote from Hollywood.*

—JdP

[149]   System Error Haiku

Note: See also "Mathematics/Poetry," item **[395]**, for more haiku.

A file that big?
It might be very useful.
But now it is gone.

Chaos reigns within.
Reflect, repent and reboot.
Order shall return.

The Web site you seek
Cannot be located
But endless others exist.

Aborted effort:
Close all that you have.
You ask far too much.

*"NO, NO, DUDLEY! WHEN I SAY I WANT TO EXECUTE
YOUR PROGRAM, THAT'S A GOOD THING!"*

First snow, then silence.
This thousand dollar screen
Dies so beautifully.

The Tao that is seen
Is not the true Tao until
You bring fresh toner.

Windows NT crashed.
Of little worth is your ire;
The network is down.

A crash reduces
Your expensive computer
To a simple stone.

Three things are certain:
Death, taxes and lost data
Guess which has occurred.

Out of memory.
We wish to hold the whole sky,
But we never will.

With searching comes loss
And the presence of absence;
"My Novel" not found.

I am the Blue Screen of Death,
No one hears your screams;
Stay the patient course.

Serious error.
All Shortcuts have disappeared.
Screen. Mind. Both are blank.

Yesterday it worked.
Today it is not working.
Windows is like that.

You step in the stream,
But the water has moved on.
This page is not here.

Having been erased,
The document you're seeking
Must now be retyped.

Rather than a beep,
Or a rude error message,
These words: file not found.

—Anonymous

[150]   Like the shoemaker's children, we have computers running out of our ears.

—Dave Farber,
Alfred Fitler Moore Professor of Telecommunication Systems,
University of Pennsylvania, Farberisms,
www.cis.upenn.edu/ farber/farism.htm.

COUNTING

[151]   If a man who cannot count finds a four-leaf clover, is he entitled to happiness?

—Stanislaw Jerzey Lec (1909–1966), Polish writer

[152]   As far as counting goes, it is said there are three kinds of people: The first kind are those who CAN count; the second kind are those who can't.

—Ron Graham*, University of California, San Diego

[153]   Whenever you can, count.

—Sir Francis Galton (1822–1911) in James R. Newman (ed.)
The World of Mathematics, New York: Simon and Schuster, 1956.

[154]   Men always want to be a woman's first love—women like to be a man's last romance.

—Oscar Wilde (1854–1900)

D

DEDUCTION VS. INDUCTION

Many dictionaries will tell you that the meanings of "deduction" and "induction" (or "inference") are all the same. To induce or to deduce, they say, is simply to argue more-or-less logically and come to a conclusion. (Test this by consulting a dictionary of your own.) The meanings of these two words may be fused in normal parlance, but statisticians will not easily replace the phrase "statistical inference" with "statistical deduction."

What's going on here? How important is it to distinguish between the two? It is interesting to note that in item [594], J. L. Heilbron takes great pains to describe how long it took before we realized what deduction was and how to use it.

You can form your own opinion. Read comments [155]–[159] of those who do see a real difference between "induction" and "deduction." Then, in items [160]–[169], share the fun as we offer definitions, examples and linguistic mishaps.

—JdP

[155]   From a drop of water, a logician could infer the possibility of an Atlantic or a Niagara without having seen or heard of one or the other. So

79

all of life is a grand chain, the nature of which is known whenever we are shown a single link of it.

—Sir Arthur Conan Doyle (1859–1930)
spoken by Sherlock Holmes in *A Study in Scarlet*

[156] ... the source of all great mathematics is the special case, the concrete example. It is frequent in mathematics that every instance of a concept of seemingly great generality is, in essence, the same as a small and concrete special case.

—Paul Halmos, *I Want to be a Mathematician*,
Washington, DC: MAA Spectrum, 1985.

[157] There is a tradition of opposition between adherents of induction and of deduction. In my view it would be just as sensible for the two ends of a worm to quarrel.

—Alfred North Whitehead (1861–1947)

[158] Emmy [Noether] was, of course, all for generalisation and I defended the relatively concrete particular cases. Then once I interrupted Emmy: "Now look here, a mathematician who can only generalise is like a monkey who can only climb UP a tree." And then Emmy broke off the discussion—she was visibly hurt.

And then I felt sorry. I didn't want to hurt anybody and especially I didn't want to hurt poor Emmy Noether. I thought about it repeatedly and finally I decided that, after all, it was not one hundred percent my fault. She should have answered, "And a mathematician who can only specialise is like a mathematician who can only climb DOWN a tree."

In fact, neither the up, nor the down, monkey is a viable creature. A real monkey must find food and escape his enemies and so he must incessantly climb up and down, up and down. A real mathematician must be able to generalise and specialise.

—George Pólya (1887–1985), from the article,
"A Story With a Moral," *Math. Gaz.* vol. 57, June 1973, p. 86.

[159] Mathematics is not a deductive science—that's a cliché. When you try to prove a theorem, you don't just list the hypotheses, and then start to reason. What you do is trial and error, experimentation, guesswork.

—Paul R. Halmos, *I Want to be a Mathematician*,
Washington, DC: MAA Spectrum, 1985.

Note: See also items **[644]–[663]** for the interplay between induction and deduction in the Michelson-Morley experiment. Item **[658]** describes the logic used to justify the conclusions on the nature of light.

[160] Definitions for Induction and Deduction

"In plain language, just tell me the difference between induction and deduction—without a lot of mumbo jumbo," Anvil Willie challenged.

We knew that deduction was more than going from the general to the particular—and that induction was more than going from the particular to the general. We, at Table Seven were not going to be a push-over audience.

Hutch addressed us all, saying, "Roughly speaking, inductive reasoning answers the question, 'what is a possible cause of what I observe?' while deductive reasoning answers the question, 'what will be the necessary consequences of what I observe?' Inductive reasoning goes to causes—deductive reasoning goes to effects.

[161]   "How about an example, then," Anvil Willie said.

"Sure," Hutch answered. "Let's say you enter a diner on a sunny day. When you finish your meal and leave, you notice that the parking lot is wet, but there is a dry spot under your car."

"So it rained while I was having my coffee, right?" Anvil Willie added hastily.

"Exactly, Willie. You just answered the question, 'what is the most likely cause for the observed dry spot?' Reasoning inductively, you infer, or induce, the existence of the rain even though you did not see it."

"What about the deduction part," Anvil Willie continued, "you know, the necessary consequence?"

"Okay. Let's stay with the car example—but suppose now you are sitting by a window of the diner and you notice it has started to rain."

"I would say that there will be a dry spot under my car when I leave the diner," Anvil Willie interrupted with obvious glee.

"Right again, Willie," Hutch said with some satisfaction. "A necessary consequence of the rain is that it will produce a dry spot under your parked car."

"Gee," Anvil Willie mused, "that car of mine is more useful than I thought—especially when it's parked."

—John de Pillis, from *Starlight Café Conversations:*
An Illustrated Dictionary from Table Seven.
(See pg. 319.)

[162] Popular Synonyms for Inductive and Deductive Reasoning

Note: See also item **[41]**, for a diagrammatic explanation of deduction and induction.

Inductive Reasoning	Deductive Result
Inference, Induction	Deduction
Cause $\Rightarrow$	Effect
Hypothesis	Result, Consequence
General Situation	Particular, Special Case
Guess	Test

Note: See also Sherlock Holmes, item **[700]** and Karl Popper, item **[171]**.

Inductive Reasoning

- **Time dependent inductive reasoning** finds a cause, causes, or conditions leading to a given observation.

  Diagnosis: of a situation is almost always an inductive process in which one seeks a past cause.

- **Non-Time dependent inductive reasoning** finds a general pattern for a given special case.

Deductive Reasoning

- **Time dependent deductive reasoning** finds an after-effect, or necessary, predicted consequence.

Testing, or verification is almost always a deductive process, seeking expected, future results.

- **Non-Time dependent deductive reasoning** finds a special case of a general situation.

Note: See also "P implies Q" = "Sufficient implies Necessary" item [20].

Examples of Induction/Deduction

[163] **The World's Funniest Joke**   Sherlock Holmes and Dr. Watson go camping, and pitch their tent under the stars. During the night, Holmes wakes his companion and says: "Watson, look up at the stars, and tell me what you deduce." Watson says, "I see millions of stars, and even if a few of these have planets like Earth out there, there might also be life."

Holmes replies: "Watson, you idiot! Somebody stole our tent!"

—Polly Stewart, Associated Press, 23 December, 2001,
in an article on the world's funniest joke of 2001
according to an Internet survey by Laughlab (laughlab.co.uk).

[164]   Watson, who deduces, is correct and Holmes, who induces (or infers), is wrong! (This is not the first time we cross swords with Mr. Sherlock Holmes as shown by the last paragraph of item [726].) Here are the details.

- Watson is *deducing* (not *inducing*), just as Holmes requested. Watson's deduction produces a predicted *result*, outcome, or effect, as per item

[162], and that is exactly what Watson properly offered when he said life was a possible effect of (the workings of) the Universe.

Watson deduces, just as Holmes requests.		
Observation as Past Cause	$\Rightarrow$	Effect
Universe observed with naked eye.	Implies, Causes, Results in	Earthlike planets that might support life.

From the Observed Universe, Watson deduces or predicts the effect, namely, existence of Life.

- Holmes, in his rebuke to Watson, is *inducing* (or *inferring*). Holmes' inference produces a past cause, or a reason for the stars being visible at that moment (see item [162]). In this case, the past cause that stars are now visible is the theft of the tent.

Holmes induces (infers), counter to his request for a deduction.		
Past Cause	$\Rightarrow$	Observation as Effect
Tent has been. stolen	Implies, Causes, Results in	Universe observed with naked eye.

From the Observed Universe, Holmes infers or induces the reason that the star-gazing is possible.

[165]

Is joke analysis proper? If we see how a joke works, do we destroy the spontaneity and fun?

Well, let's see. When you find out how a magic trick works, are you disappointed? Are you saddened when film makers tell you how they created their special effects? Does the cosmos become less interesting once you understand Special Relativity's explanation of time dilation and distance-contraction?

This joke is funny and it also provides a good illustration of the difference between induction and deduction. It is multi-faceted.

*So when we ask whether joke analysis is proper, we might
also ask: How can knowing more turn out to be less of an
experience?*

—JdP

Note: Real life scientists can confuse "induction" with "deduction." See
item **[169]** for a case in point.

[166] Smoke and Fire. The well-known expression, "where there is
smoke, there is fire," can be restated as, "you can be sure that fire is always
one cause of smoke." From observed smoke, fire is inferred, or induced as
a past cause.

Cause	⇒	Predicted effect
Fire, moisture	Imply, cause	Smoke

To go in the other direction, suppose we observe dead leaves with
moisture. Once we set fire to it, we would predict (deduce) a future effect, namely, that the fire's heat would eventually transform the moisture
into visible steam or smoke.

[167] Inferring the Origin of the Cosmos: the Big Bang In one sense,
the inference that a Big Bang started (caused) the Universe is obvious. After
all, if galaxies are flying away from each other everywhere in the Universe,
like dots on the surface of an expanding balloon, then by playing the tape
backwards, there must have been a single point and a Big Bang from which
space, time, matter and all the Universe, originated.

But how obvious is the Big Bang idea?

Some History: Red shifts of light from galaxies, signs that galaxies are
receding from the observer, were first observed in 1917 by American astronomer, Vesto Melvin Slipher (1875–1969). (Red shifts are result of the
Doppler effect which says: Just as the sound of whistle of a receding train
takes on a lower frequency, *i.e.*, a lower pitch, so does the light of a receding star take on a lower frequency, *i.e.*, we see a shift toward the lower
frequency red.) But Slipher did not know the distances involved so no coherent structure could be inferred from these receding galaxies. Red shifts
alone were not enough to warrant support for the Big Bang inference.

Then, twelve years later in 1929, Edwin Powell Hubble (1889–1953) measured and plotted those missing distances against velocities relative to the Earth. Among the twenty-five galaxies, he found a linear straight-line relationship. (Galaxies twice the distance from Earth as another galaxy race away from Earth at twice the speed of that galaxy; at three times the distance, three times the speed, etc.) This could only mean that galaxies were, indeed, moving away from each other like spots on an expanding balloon—the Universe was expanding with some galaxies fleeing from Earth at two million miles per hour.

Yet Hubble was without a theory for this linear structure and this made him apprehensive. (Einstein's theory of General Relativity, circa 1915, did predict an expanding Universe. But even Einstein's intuition left him, at first, unwilling to accept the idea of an expanding Universe. Hubble was totally unaware of Einstein's theoretical work.) Instead of using the term "expanding Universe," Hubble spoke of "redshift-distance relation" or "velocity shifts." It is 1929 and there is still no wide support for the Big Bang.

The English astrophysicist, Fred Hoyle (1915–2001), was championing the alternate theory that the Universe evolved in a steady state from materials that already existed. He coined the term "Big Bang," meaning it to be derisive. This was 1950 and still, not many scientists are warming to the Big Bang theory.

In 1948, physicist Georg Gamov (1904–1968) noted that if a hot burst of the Big Bang had occurred, then one could predict (deduce) that huge amounts of radiation would have resulted and, in a cooling Universe, the residue of that radiation could be measured even today.

It took until 1965 for the measurement of Gamov's Cosmic Background radiation to be made. In so doing, Arno Penzias and Robert Wilson of Bell Labs provided the strongest evidence of the Big Bang to date. They were awarded the Nobel Prize in Physics in 1978.

That, briefly, is the story of how an expanding Universe led to the inference (cause) of a Big Bang that took some fifty years to become accepted.

(For more on the Big Bang, see Timothy Ferris' book, *Coming of Age in the Milky Way*, Anchor Books, 1988, ch. 11.)

[168]   Inferring Why Dinosaurs Went Extinct

Some Background: Wolfgang Amadeus Mozart (1756–1791) and inventor Benjamin Franklin (1706–1790) never heard of them. Our knowledge of chocolate (introduced in London in 1657) goes back much further than our knowledge of these creatures. It was not until 1841 that the word "dinosaur," from the Greek "deinos" meaning fearfully great, and "sauros" meaning lizard, was introduced by Sir Richard Owen (1804–1892). Before then, there was no concept of anything like a dinosaur.

For almost two centuries, we have been asking: Why did the dinosaurs go extinct? The inferred cause is that a meteorite impact did it. Here is how we finally came to accept this inference with a high degree of confidence.

Some History: First, Luis and Walter Alvarez (father and son) in the late 1970's, in Gubbio, Italy found a layer of clay with thirty times the normal concentration of the rare element iridium. Iridium is not so rare in meteors. The Alvarez clay was dated as 65 million years old.

Then, in 1990, Alan Hildebrand, while searching for oil in Mexico, found evidence of a meteorite impact off the northwest tip of the Yucatan Peninsula. The size of the crater was consistent with one that might be caused by a meteor containing enough iridium to cause the very concentration measured by the Alvarez team. It was soon determined that Hildebrand's crater contained the anticipated concentration of iridium. The crater was dated as 65 million years old.

This is how several sources (age dating, iridium concentration, temperature patterns), not just one source, gave us a foundation for confidence in the inference that a meteor impact caused the extinction of the dinosaurs.

Notes. See item [72] for Michael Shermer's Baloney Detection Rule no. 6 that states the importance in science to have several sources supporting a single conclusion.

See also item [551] for Stephen J. Gould's comments on the amazing survival of planetary life.

[169] Yes, Even Scientists Can Be Confused

Alas, at times, scientists will confuse "induction" with "deduction." For example, in a Public Broadcasting Service promotional announcement for NOVA's science program, Neanderthals on Trial, (broadcast Jan. 22, 2002), there is the statement,

> *"Dig and Deduce (Hot Science): How can old bones tell us how ancient people lived?"*

Dig and Deduce? Don't they mean "Dig and Induce" or "Dig and Infer?" Let's look into this further.

During the program, an archeologist asks whether certain patterns of animal bones found in a cave are

*(a) **caused by** humans disposing of a cooked meal, or*

*(b) **caused by** the bones being deposited by a flowing stream of water.*

*The archeologist is seeking an event that causes the observed pattern of bones. As item [162] tells us, whenever we seek **causes** of an observation (such as bone patterns) then we are inducing or inferring, not deducing or predicting.*

This case is typical of inference in the sense that more than one possible cause of the observed event is put on the table. In our case, two candidate causes, (a) and (b) above, are given. On the other hand, there is much less wiggle room in deduction, or implication, or seeking effects. Deduction is often more precise than induction or inference.

Yes, there will be worse problems in the world than the confusion of "induction" with "deduction." This assertion will be given due consideration once I can determine whether I inferred it or deduced it!

—JdP

Note: On the distinction between "induction" and "deduction," even Sherlock Holmes can become confused, as items [163]–[164] reveal. On the other hand, as we see in items [156]–[159], and [594], some take the difference very seriously.

Popper's Falsifiability Test in Scientific Method

[170]   The basic method of science as identified by Karl Popper—

make a guess and then test it

—is the essence of all thinking. You do it, I do it, cats do it, even worms do it. For day-to-day purposes, there is no other way to get a feel for whatever is going on. Seen in this light, science emerges as the most natural process of all. The unnaturalness (if such it is) of science lies only in its explicitness: that it lays out problems for inspection, while our own common-sensical brains, bent on survival, draw lightning conclusions from fleeting impressions and are content with imperfection, provided it works.

—Colin Tudge, "Why Science Should Warm Our Hearts,"
New Statesman, Feb. 26, 2001.

[171] **Falsifiability:**   Karl Popper's principal contribution to the philosophy of science rests on his rejection of the inductive method in the empirical sciences. According to this traditional view, a scientific hypothesis may be tested and verified by obtaining the repeated outcome of substantiating observations. As the Scottish empiricist David Hume had shown, however, only an infinite number of such confirming results could prove the theory correct. Popper argued, instead, that hypotheses are deductively validated by what he called the "falsifiability criterion." [That is, one] seeks to discover an observed exception to [a guessed-at] rule. The absence of contradictory evidence thereby becomes corroboration of his theory. According to Popper, such pseudosciences as astrology, metaphysics, Marxist history, and Freudian psychoanalysis are not empirical sciences, because of their failure to adhere to the principle of falsifiability.

—*Encyclopaedia Britannica*,
"The Philosophy of Karl R. Popper," (1902–1994)

Note: See Edsger W. Dijkstra, item [143], who applies Karl Popper's principle to computer programming, and David Finkelstein, item [676], who comments on the growth of physics through violation and modification of its laws.

Popper's falsifiability can be regarded as a physicist's use of the mathematical technique of reductio ad absurdum or proof by contradiction. This centuries-old tool of logic had been used by Euclid. (See item [598] for a definition and items [599]–[602] for an application in mathematics. See also item [660] for more on the connection between falsifiability, modus ponens and proof by contradiction.

Aristotle and Testing

[172]

Is testing an obvious thing to do? Scientific theories were often declared to be true without regard to verification as described in **[170]**. *Testing was not advocated by Aristotle (384–322* B.C.*), a student of Plato and tutor of Alexander the Great. For example, Aristotle inferred or guessed that matter was composed of four basic elements, earth, air, fire and water. He even asserted (but never tested) that women have fewer teeth than men do. (See Aristotle's* History of Animals, Book II, part 3.*)*

Why did ancient scholars buy into Aristotle's untested structure? Part of the reason is that Truth was accepted by the power of Authority (of which Aristotle was one) and also, the concept of testing, deducing and predicting was not part of the scientific thinking at the time. See Heilbron, item **[594]** *to see how the art of deduction was a long time in coming.*

—JdP

Wittgenstein vs. Popper: The Poker Debate

[173]

Should testing make you angry? If, for example, an objective test is made to decide whether a bucket leaks, or whether an unboiled egg will spin on a table, then why should emotions come into play at all? The following story shows how this can happen.

Philosophers Karl Popper and Ludwig Wittgenstein, both Austrian expatriates, held different views on the validity of testing a presumed external reality. They met in 1946 at Cambridge University's Moral Science Club with an audience of philosophers and students to publicly discuss their views.

As [170] indicates, Popper held that there is an external, testable reality "out there." (See [673] for related concept of Platonism.) On the other hand, Wittgenstein saw reality as internal—mere constructs of the mind or individual interpretations of our sensory inputs. Language, Wittgenstein believed, is only a self-referential game that therefore cannot be used to probe or understand the world.

This distinction is important because it was Wittgenstein's philosophy that eventually led to the ideas of Thomas Kuhn. It is one thing to say we cannot know what is "out there" and quite another (as the postmodernists have done) to add that whatever you decide is true "out there" is okay with me. (See items [87]–[90] for postmodern assertions that definitions need not be precise, items [91]–[96] on Alan Sokal's hoax, and items [580]–[584] for Kuhn's disagreement with scientists on the nature and progress of science.)

Therefore, universally valid, clear definitions make no sense since your definition is as valid as mine. Since language is presumed to be interpreted and constructed differently for each of us, seeking a single standard of clarity becomes a ridiculous notion.

Getting back to the Cambridge debate: Wittgenstein stormed out of the debate after ten minutes, having been made so angry that he actually brandished a fireplace poker at Popper. Eye-witness accounts differ but in answer to Wittgenstein's challenge, asking Popper to state at least one "externally valid" moral principle, Popper answered by saying something like, "Not to threaten visiting lecturers with pokers."

—JdP

Note: For a lively account of this dispute, see *Wittgenstein's Poker: The Story of a Ten-Minute Argument Between Two Great Philosophers*, by David Edmonds, John Eidinow, Ecco Press, 2002. For more on Wittgenstein and his denial that there exists a shared (or absolute) truth, see items [105], [106].

Reliability of Eye-Witness Accounts

[174]

Seeing is Believing? *Eyewitness accounts of the ten-minute episode between Ludwig Wittgenstein and Karl Popper seem anything but reliable. Imagine! The Cambridge audience consisted of not-stupid people (philosophers, students) and they still could not agree on what they saw happen at that Moral Science Club meeting in 1946. As we have seen in items* **[97]** *and* **[580]**, *it's bad enough to live with the unreliability of second-hand information (when A tells B what C is thinking), but what do we make of unreliability of first-hand information (when A tells B what A is thinking)?*

Is this enough to make you rub your own eyes in disbelief at the unreliability of eyewitnesses? Or to throw up your hands and say, oh well, I suppose Wittgenstein had it right—there is no knowable reality "out there" that any two people will ever agree on. Or is Wittgenstein, along with the post-modernists and moral relativists, giving themselves a free pass, excusing themselves from the difficult job (responsibility) of figuring it all out? (See item **[687]** *for the comment of Henry James on absolute truth.)*

—JdP

DIMENSION

Note: Items **[617]**–**[625]** show how higher dimensions can be understood even though they exceed our ability to model them.

[175] A conversational defense every mathematician needs at parties: How do you reply when someone asks, "What is the fourth dimension?" I usually say, "I'll point in that direction, if you point in the first."

—Tim Poston*, Chief Scientist,
Digital Medicine Lab, Johns Hopkins, Singapore

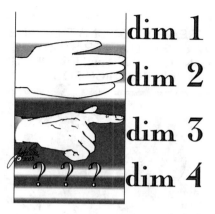

[176] One theme that comes up in many of my lectures is "The Fourth Dimension: It's not just 'time' anymore." To illustrate, I tell the story of the herb distributor who wanted to use a vector space to keep track of orders. There would be one coordinate for parsley, one for sage, one for rosemary, and one for oregano—because the fourth dimension isn't just thyme anymore. The reaction of the class? Well, let us just say their groans give my story yet *another* dimension.

—Tom Banchoff*, Brown University

[177] "Look, we live in three dimensions. Why on earth would mathematicians need to study more dimensions than that?" Anvil Willie challenged.

"Are you ready for this?" Hutch replied in a calm and measured tone. "In my computer graphics class, I need four-dimensional ideas to produce those great three-dimensional effects that you see on a two-dimensional movie screen."

"You need four dimensions to create pictures of three-dimensional mountains, caverns and cars that I look at on a two-dimensional screen?"

Hutch answered with definiteness, "Correct! We have to embed three-dimensional objects, like mountains and cars, into four dimensions so we can use an important mathematical property called linearity."

"Huh?"

"Linearity. It simplifies calculations," Hutch added. "Linearity is what lets us produce the movements of objects (stretching, rotation, translating) with something called matrices. And we need a full four dimensions to use these matrices."

GREAT QUESTIONS THAT ARE
ANSWERED AFTER STUDYING
the INHABITANTS and PROPERTIES
of DIMENSION FOUR.

LOVELY NEW HAT!
IS IT FROM A
3-D MAYTAG OR
A KENMORE?

• *Where does the missing sock from your laundry go?*

"I see. I suppose you really do need four dimensions to create graphics effects." Anvil Willie exhaled in a long breath of resignation as he stirred his coffee. "All right, Hutch. I'll give you four dimensions. But let's not talk about five or six dimensions until next week at least, okay?"

—John de Pillis from *Starlight Café Conversations:
An Illustrated Dictionary from Table Seven.* (See pg. 319.)

DISCOVERY

Note: See item [584] for Max Planck's comments on innovation.

[178]   What you have been obliged to discover by yourself leaves a path in your mind which you can use again when the need arises.

—G. C. Lichtenberg

[179]   I hope that posterity will judge me kindly, not only as to the things which I have explained, but also to those which I have intentionally omitted so as to leave to others the pleasure of discovery.

—René Descartes (1596–1650) *La Géométrie*

[180]   In a sense, it was a Goldilocks experience: On the stable manifold everything moves exponentially to the origin, and on the unstable manifold everything moves exponentially away from the origin. The manifold that was just right—that was the center manifold!

> —Al Kelley*, University of California Santa Cruz, on how he developed the idea of the center manifold in the theory of ordinary differential equations.

[181]   If I found any new truths in the sciences, I can say that they follow from, or depend on, five or six principal problems which I succeeded in solving and which I regard as so many battles where the fortunes of war were on my side.

> —René Descartes (1596–1650) *Discours de la Méthode*, 1637.

[182]   The first seven years I'd worked on this problem [the Taniyama-Shimura conjecture that would prove Fermat's Last Theorem as a consequence], I loved every minute of it. However hard it had been, there'd been setbacks often, there'd been things that had seemed insurmountable, but it was a kind of private and very personal battle I was engaged in.

> —Andrew Wiles, from the BBC program, 15 January 1996, *Fermat's Last Theorem* (broadcast in the U.S. on PBS's NOVA as *The Proof*), by Simon Singh and John Lynch.

Note: See also Joel Spencer, item [348], observing non-mathematicians who prefer to accept as reality, only the stereotypical, looney research mathematician.

E

EINSTEIN

[183] When Einstein first came to the Institute of Advanced Study at Princeton a large crowd attended his first lecture. He commented "I never realized that in America, there was so much interest in tensor analysis."

—quoted in *Mathematical Reminiscences*,
by Howard Eves, Math. Assoc. Amer. 2001.

[184]

So you thought it was just dots...?

The scientist is immediately struck by the way Einstein has utilized the various achievements in physics and mathematics to build up a coordinated system showing connecting links where heretofore none was perceived. The philosopher is equally fascinated by a theory which, in detail extremely complex, shows a singular beauty when viewed as a whole.

—B. Harrow

[185] At any rate, I am convinced that He [God] does not play dice.

—Albert Einstein (1879–1955), from *Einstein und Born Briefwechsel* (1969) pg. 130 (stated as "Gott würfelt nicht." in a letter to Max Born, 4 Dec. 1926.) This quote sometimes takes the form, "God does not play dice with the universe."

[186]

Einstein and the Ice Cream Cone
His first day at Princeton, the legend goes
    he went for a stroll (in his rumpled clothes).
He entered a coffeeshop—moment of doubt—
    then climbed on a stool and looked about.
Beside him a frosh, likewise strange and alone,
    consoling himself with an ice-cream cone.
Now Einstein's glee was plain to see
    At the sight of a CONE with a SPHERE on top
    (in the hand of a frosh in a sandwich shop)
And—oh incredible—
    completely edible!
He smiled at the frosh, then the waiter came,
    and Einstein gestured he'd like the same,
And they sat there nibbling, suddenly kin,
    with no common language to verbalize in,
But foreign no longer, no longer alone,
    with the fellowship bond of an ice-cream cone.
—Katherine O'Brien, *The Mathematics Teacher*, April 1968.

ELEGANCE

[187]   What is it indeed that gives us the feeling of elegance in a solution, in demonstration? It is the harmony of the diverse parts, their symmetry, their happy balance; in a word it is all that introduces order, all that gives unity, that permits us to see clearly and to comprehend at once both the ensemble and the details.

—Henri Poincaré (1854–1912)

[188]   On its surface, mathematics appears to be the most objective of sciences—results are either true or they are false. But at its higher levels, mathematics becomes an art—results become "beautiful," "elegant," or "clean." Mathematics becomes a matter of taste.

—Lisette de Pillis, Harvey Mudd College

[189] Occam's Razor (Ockham's Razor)

"*Entia non sunt multiplicanda praeter necessititem.*" Translated as, "entities are not to be multiplied beyond necessity." Called The Principle of Parsimony or the Law of Economy, Occam's Razor essentially means that if two explanations equally explain the facts, the one with the fewer postulates shall be chosen.

—William of Occam (1285–1349)

[190]

According to the Encyclopaedia Britannica,* the principle of Ockham's Razor was, in fact, invoked before Ockham by Durand de Saint-Pourçain (1270–1334), a French Dominican theologian and philosopher of "dubious orthodoxy."

Afterwards, Nicole d'Oresme (1325–1382), a 14th-century French physicist, invoked the law of economy, as did Galileo (1564–1642), who defended simplicity in describing the workings of the heavens.)

Yet, because Ockham mentioned the principle so frequently and employed it so sharply, his is the name that is attached to this principle of parsimony. One example of Ockham's use is in his "simple" description of motion as merely the reappearance of a thing in a different place.

—JdP

*Encyclopedia Britannica online, http://www.britannica.com/eb/article?eu=58133&tocid=0

Note: See also Albert Einstein's comment on simplicity, item **[541]**.

EPSILON

[191]

Many epsilons doth a delta make.

—John de Pillis

Or, to put it another way,

[192]   No snowflake in an avalanche ever feels responsible.
—Stanislaw Jerzey Lec (1909–1966), Polish writer

Pensieri spettinati, Bompiani, Milano, 1984 [1965], Nuovo portico

[193] There's A Delta For Every Epsilon (Calypso)

(Tom Lehrer Lyrics) To be sung to calypso music.

> There's a delta for every epsilon,
> It's a fact that you can always count upon.
> There's a delta for every epsilon
> > And now and again,
> > There's also an N.
>
> But one condition I must give:
> The epsilon must be positive
> A lonely life all the others live,
> > In no theorem
> > A delta for them.
>
> How sad, how cruel, how tragic,
> How pitiful, and other adjec-
> Tives that I might mention.
> The matter merits our attention.
> If an epsilon is a hero,

Just because it is greater than zero,
It must be mighty discouragin'
To lie to the left of the origin.

This rank discrimination is not for us,
We must fight for an enlightened calculus,
Where epsilons all, both minus and plus,
    Have deltas
    To call their own.

        —Tom Lehrer, *The American Mathematical Monthly*,
        vol. 81, 1974, p. 612.

EQUATIONS

Note: See also item **[661]** for comments on Maxwell's Equations.

[194] It is often said that an equation contains only what has been put into it. It is easy to reply that the new form under which things are found often constitutes by itself an important discovery. But there is something more: analysis, by the simple play of its symbols, may suggest generalizations far beyond the original limits.

        —Emile Picard (1856–1941)

[195] ...from the time of Kepler to that of Newton, and from Newton to Hartley, not only all things in external nature, but the subtlest mysteries of life and organization, and even of the intellect and moral being, were conjured within the magic circle of mathematical formulae.

        —Samuel Taylor Coleridge (1772–1834),
        The Theory of Life.

EUCLID

Note: See also R. L. Heilbron, item **[594]**, for a discussion of the development of the deductive process. For the first recorded records of Euclid's formal definitions, see *The Forgotten Revolution*, item **[589]**.

[196] According to legend, someone who began to read geometry with Euclid had no sooner learned the first theorem, when he asked "What shall I get by learning these things?" Whereupon Euclid called an assistant and said "Give him three-pence, since he must make a gain out of what he learns."

—Timothy Ferris, *Coming of Age in the Milky Way*,
Anchor Books, 1988, p. 41

Note: See also items **[415]** and **[416]** on Usefulness.

[197]

In the Greek mathematical Forum
Young Euclid was present to bore 'em.
    He spent most of his time
    Drawing circles sublime
And crossing the pons asinorum.

—Leo Moser, *The American Mathematical Monthly*,
vol. 80, no. 8, 1973, p. 902.

I love my wife, but Oh, Euclid!

[198] My model is Euclid, whose justly celebrated book of short stories, entitled *The Elements of Geometry* will live when most of us who are scribbling today are forgotten. Euclid lays down his plot, sets instantly to work at its development, letting no incident creep in that does not bear relation to the climax, using no unnecessary word, always keeping his one end in view, and the moment he reaches the culmination he stops.

—Robert Barr (1850–1912) from his article, "How to Write
a Short Story," in *The Bookman A Literary Journal*,
vol. 5, March, 1897. No. 1.

EULER

[199]   Euler calculated without effort, just as men breathe, as eagles sustain themselves in the air.

—François Arago (1786–1853) from *Comic sections: the book of mathematical jokes, humour, wit, and wisdom,* by Desmond MacHale, Dublin 1993.

[200]   It is the invaluable merit of the great Basel mathematician Leonhard Euler, to have freed the analytical calculus from all geometric bounds, and thus to have established analysis as an independent science, which from his time on has maintained an unchallenged leadership in the field of mathematics.

—Thomas Reid

[201]   Euler—the unsurpassed master of analytic invention.

—Richard Courant (1888–1972)

[202]   Euler who could have been called, almost without metaphor, and certainly without hyperbole, analysis incarnate.

—François Arago (1786–1853)

EXPERIMENTS

(Item [214] offers Martin Gardner's observations on theory without experimental confirmation.)

[203]   Erasmus Darwin* had a theory that once in a while one should perform a damn-fool experiment. It almost always fails, but when it does come off, it is terrific. Darwin played the trombone to his tulips. The result of this particular experiment was negative.

—J. E. Littlewood (1885–1977), *A mathematician's miscellany*

*Note: Darwin was grandfather of both the naturalist Charles Darwin and the biologist Francis Galton.

F

FAITH, RELIGION, AND SPIRITUALITY

[204]   I believe in the brotherhood of man and the uniqueness of the individual. But if you ask me to prove what I believe, I can't. You know them to be true but you could spend a whole lifetime without being able to prove them. The mind can proceed only so far upon what it knows and can prove. There comes a point where the mind takes a higher plane of knowledge, but can never prove how it got there. All great discoveries have involved such a leap.

> —Albert Einstein (1879–1955) From *Einstein: The Life and Times*,
> by Ronald W. Clark, p. 622, Avon, Reissue edition (1999).

Note: See also Kurt Gödel's Incompleteness Theorem, item [324], for a statement on the limits of all logical systems.

[205]   ... There can be no doubt about faith and not reason being the ultimate ratio. Even Euclid, who has laid himself as little open to the charge of credulity as any writer who ever lived, cannot get beyond this. He has no demonstrable first premise. He requires postulates and axioms which transcend demonstration, and without which he can do nothing. His superstructure indeed is demonstration, but his ground is faith. Nor again can he get further than telling a man he is a fool if he persists in differing from him. He says "which is absurd, " and declines to discuss the matter further. Faith and authority, therefore, prove to be as necessary for him as for anyone else.

> —Samuel Butler (1612–1680), *The Way of All Flesh*.

Note: See also "Axioms and Definition", item **[68]**.

[206]   There is more religion in men's science than there is science in their religion.

—David Henry Thoreau (1817–1862), from
A Week on the Concord and Merrimack Rivers.

[207]   In his book, *The Whys of a Philosophical Scrivener*, Martin Gardner defines and defends *fideism* at length. It is a pragmatic argument, taken from the philosophers William James, Charles Pierce, and Miguel Unamuno. At its core, [*fideism* is defined by:]

[If]

(1) in issues of extreme importance to human existence,

[and]

(2) when the evidence is inconclusive one way or the other,

[and]

(3) you must make a choice,

[then]

it is acceptable to take a leap of faith. Martin Gardner, the skeptic of all skeptics, is a fideist.

—Michael Shermer, from *How We Believe*,
W. H. Freeman and Company, New York, 2000, p. 97,
on the faith of skeptic Martin Gardner.

Note: Martin Gardner is not the only one confronted with a toggling, either-or situation. See also Abraham Lincoln, items **[274]**, **[275]**; Eric Schechter, item **[340]**; the Bible, item **[269]**; and finally, for the role of toggling in the Monty Hall problem, see item **[300]**.

[208]   There are problems to whose solution I would attach an infinitely greater importance than to those of mathematics, for example touching ethics, or our relation to God, or concerning our destiny and our future; but their solution lies wholly beyond us and completely outside the province of science.

—Carl Friedrich Gauss (1777–1855), In James R. Newman (ed.) *The World of Mathematics*, New York: Simon and Schuster, 1956, p. 314.

[209]   No longer is theology embarrassed by the contradiction between God's immanence and transcendence. Hyperspace touches every point of three-space. God is closer to us than our breathing. He can see every portion of our world, touch every particle without moving a finger through our space. Yet the Kingdom of God is completely "outside" of three-space, in a direction in which we cannot even point.

—Martin Gardner, science writer, quoted in *The Fourth Dimension: A Guided Tour of the Higher Universes*, by Rudy Rucker, Houghton Mifflin Co., 1985.

Note: See also Breaking the Bonds of Dimension, item **[617]**, which explores concepts beyond familiar three-dimensional space.

[210]   Science and religion are both extremely important issues, but they do different things. Science deals with the factual state of nature, and religion deals with ethics and meaning.

—Stephen J. Gould, from "Questioning the Millennium: Why We Cannot Predict the Future," Inaugural Farfel Distinguished Lecture, University of Houston, Nov. 6, 2000.

[211] Every formula which expresses a law of nature is a hymn of praise to God.

> —Maria Mitchell (1818–1889), astronomer, first woman elected to the American Academy of Arts and Sciences (1848). Quoted as inscription on her bust in the Hall of Fame (1905).

[212] I feel most deeply that the whole subject is too profound for the human intellect. A dog might as well speculate on the mind of Newton. Let each man hope and believe what he can.

> —Charles Darwin (1809–1882), From a letter to American botanist Asa Gray (1860) on whether human reason could ever explicate God. (For limitations of the human mind, see Ron Graham, item **[268]** and item **[324]** for Kurt Gödel's incompleteness theorem stating that some true statements can never be proven.)

Note: For more on the exchange between Asa Gray and Charles Darwin as he wrestled with his changing views of God, see *Did Adam and Eve Have Navels?* by Martin Gardner, W. W. Norton & Co., 2000, pp. 22, 167.

[213] No child under the age of fifteen should receive instruction in subjects which may possibly be the vehicle of serious error, such as philosophy or religion, for wrong notions imbibed early can seldom be rooted out, and of all the intellectual faculties, judgment is the last to arrive at maturity. The child should give its attention either to subjects where no error is possible at all, such as mathematics, or to those in which there is no particular danger in making a mistake, such as languages, natural science, history, and so on.

—Arthur Schopenhauer (1788–1860) *On Education* (1851).

Note: See also George Bernard Shaw, item **[688]** on the inerrancy of religion and the fallibility of science.

[214] The stark truth is that there is not the slightest shred of reliable evidence that there is any universe other than the one we are in. No multiverse theory has so far provided a prediction that can be tested. In my layman's opinion they are all frivolous fantasies. As far as we can tell, universes are not as plentiful as even two blackberries. Surely the conjecture that there is just one universe and its Creator is infinitely simpler and easier to believe than that there are countless billions upon billions of worlds, constantly increasing in number and created by nobody. I can only marvel at the low state to which today's philosophy of science has fallen.

—Martin Gardner, science writer from his article, "Notes of a Fringe-Watcher: Multiverses and Blackberries" in *The Skeptical Inquirer*, vol. 25, no. 4, July–August 2001.

[215] Everyone who is seriously involved in the pursuit of science becomes convinced that a spirit is manifest in the laws of the Universe—a spirit vastly superior to that of man.... In this way the pursuit of science leads to a religious feeling of a special sort, which is indeed quite different from the religiosity of someone more naive.

—Albert Einstein, (1879–1955), Letter to a child who asked if scientists pray, January 24, 1936; Einstein Archive 42-601.

[216] Science without religion is lame; religion without science is blind.

—Albert Einstein (1879–1955), from *Subtle Is the Lord: The Science and the Life of Albert Einstein*, by Abraham Pais, Oxford University Press, Oxford & New York, 1982.

FORMULAS

[217] One cannot escape the feeling that these mathematical formulas have an independent existence and intelligence of their own, that they are

wiser than we are, wiser even than their discoverers, that we get more out of them than was originally put into them.

—Heinrich Hertz (1857–1894) quoted by Morris Kline in
Mathematics and the Search for Knowledge,
section on Maxwell's Equations, p. 144.

Note: See also item **[661]** for more on Maxwell's equations.

[218]   Some textbooks leave the quantity 0^0 undefined, because the functions x^0 and 0^x have different limiting values when x decreases to 0. But this is a mistake. We must define

$$x^0 = 1 \qquad \text{for all } x,$$

if the binomial theorem is to be valid when $x = 0$, $y = 0$, and/or $x = -y$. The theorem is too important to be arbitrarily restricted! By contrast, the function 0^x is quite unimportant.

—Ron Graham, Don Knuth, Orin Patashnik, *Concrete Mathematics*,
p. 162, Addison-Wesley, 2nd printing Dec., 1988.

FOUR COLOR PROBLEM

[219]

The Four Color conjecture describes a property enjoyed by any flat (planar) map of countries, each of which is a single, connected region. (This excludes two-part

partitioned regions like the state of Michigan.) The
property is this: such a planar map can be colored with no
more than four colors in such a way that adjacent
countries always have different colors. This statement was
first articulated in 1852 when Francis Guthrie, while
trying to color the map of counties of England (where
adjacent regions were assigned different colors), noticed
that four colors sufficed. He asked his brother, Frederick,
whether any map could be so colored using at most four
colors. Frederick Guthrie brought the question to
Augustus De Morgan (1806–1871). The first printed
reference is due to Arthur Cayley (1821–1895), "On the
colourings of maps." Proc. Royal Geog. Soc. 1 (1879),
259–261.

This theorem was first proved in 1976 by Kenneth Appel
and Wolfgang Haken who programmed computers to color
1, 936 basic forms of maps with four colors. From this
they deduced that any given map can be colored with at
most four colors.

To date, nobody has been able to prove this theorem
without using a computer although, in principle, it is
possible to prove it by hand—possibly before the sun
cools. With Appel and Haken's result, the very nature of
proof becomes a topic of heated debate. In the absence of
a proof produced entirely by a human being, should
mathematicians take the word of a machine—"a bucket of
bolts?" So far, the score is "bucket 1, humans 0."

—JdP

Note: See also item **[606]** for the role of probability and the computer in testing suspected prime numbers.

[220]   So Appel and Haken solved the Four Color Problem with a computer proof in 1976. But they forgot one thing—to tell us what those four colors are.

—Lew Lefton, Georgia Institute of Technology

FRUSTRATION AND PLEASURE

[221]   I love mathematics—where else can you get your fill of all the bits and bytes you want and not gain any weight at all?

—Burt Nanus*, University of Southern California

[222]   Mathematicians grow very old; it is a healthy profession. The reason you live long is that you have pleasant thoughts. Math and physics are very pleasant things to do.

—Dirk Jan Struik (1894–2000)
MAA Focus, vol. 20, no. 9, pg. 16, Dec 2000.

[223]   My life is spent in one long effort to escape from the commonplaces of existence. These little problems help me to do so.

—Sir Arthur Conan Doyle (1859–1930)
spoken by Sherlock Holmes in *The Red-Headed League*

"I'm a mathematician. My problems are both real and complex."

[224]   It almost looks as if analysis were the third of those impossible professions in which one can be sure beforehand of achieving unsatisfactory results. The other two, which have been known much longer, are education and government.

—Sigmund Freud, (1856–1939) *On Freud's Analysis Terminable and Interminable*, Yale University Press, Joseph Sandler, ed., 1991, originally published 1937.

[225] I still say to myself when I am depressed, and find myself forced to listen to pompous and tiresome people, "Well, I have done one thing you could never have done, and that is to have collaborated with both Littlewood and Ramanujan on something like equal terms."

—Godfrey H. Hardy (1877–1947) *A Mathematician's Apology*, London, Cambridge University Press, 1941.

[226] Don't be intimidated! I have seen many people get discouraged because they see mathematics as full of deep incomprehensible theories.

There is no reason to feel that way.

In mathematics whatever you learn is yours and you build it up—one step at a time. It's not like a real time game of winning and losing. You win if you are benefitted from the power, rigor and beauty of mathematics. It is a big win if you discover a new principle or solve a tough problem.

—Fan Chung, University of California, San Diego *Math Horizons*, Sept. 1995, pp. 14–18, from an interview with Don Albers in which she offers advice to young women considering careers in mathematics.

G

GEOMETRY

Note: See V. I. Arnold, item **[12]** for thoughts on the importance of geometry relative to the real world.

[227] All geometrical reasoning is, in the last resort, circular: if we start by assuming points, they can only be defined by the lines or planes which relate them; and if we start by assuming lines or planes, they can only be defined by the points through which they pass.

—Bertrand Russell (1872–1970)
An Essay on the Foundations of Geometry, Ch. III, p. 54.

[228] Geometry is the science of correct reasoning on incorrect figures.

—George Pólya (1887–1985)
How to Solve It, 1945, pg. 75.

[229]

Pólya's Irony: In item [228], Pólya is saying that correct reasoning can be effective even on hand-drawn, imprecise diagrams. The following diagram shows that correct reasoning on an imprecise diagram can also lead us astray.

Although one square inch (dark area) is removed from the 110 inch squared figure at the left-hand side of the diagram, the total area is not diminished, as the figure on the right-hand side illustrates. (The incorrectness of the figure is in the "apparent" isosceles 1 in. by 1 in. triangles.)

—JdP

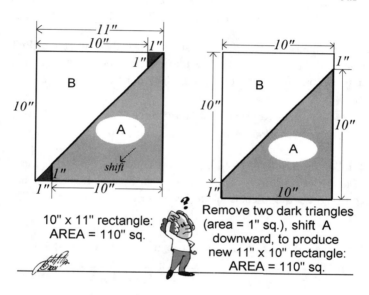

10" x 11" rectangle:
AREA = 110" sq.

Remove two dark triangles (area = 1" sq.), shift A downward, to produce new 11" x 10" rectangle: AREA = 110" sq.

[230] Arithmetic symbols are written diagrams and geometrical figures are graphic formulas.

—David Hilbert (1862–1943)
Amer. Math. Soc. Bulletin, 1902, vol. 8, pg. 443.

[231] Projective geometry has opened up for us with the greatest facility, new territories in our science, and has rightly been called a royal road to its own particular field of knowledge.

—Felix Klein (1849–1945) quoted in E. T. Bell's *Men of Mathematics*, New York: Simon and Schuster, pg. 206, 1937.

[232] I have discovered such wonderful things that I was amazed...out of nothing I have created a strange new universe.

—János Bolyai, (1802–1860) From an 1823 letter to his father, in reference to the creation of a non-euclidean geometry.

[233] For God's sake, please give it up. Fear it no less than the sensual passion, because it, too, may take up all your time and deprive you of your health, peace of mind and happiness in life.

—Wolfgang Bolyai, (1775–1856) [To son János Bolyai, urging him to give up work on non-Euclidian geometry.]

Philip J. Davis and Reuben Hersh, *The Mathematical Experience*, Boston: Houghton Mifflin Co., 1981, pg. 220.

[234] Geometry has two great treasures: one is the Theorem of Pythagoras; the other, the division of a line into extreme and mean ratio. The first we may compare to a measure of gold. The second we may name a precious jewel.

—Johannes Kepler (1571–1630) quoted Carl B. Boyer's *A History of Mathematics*, 2nd ed., John Wiley & Sons, April 1989.

Note: See also items **[611]**–**[655]** for applications of the Pythagorean Theorem in the explanations of time dilation in Special Relativity, the fixed speed of light in the Michelson-Morley experiment, and efficiency in pizza cutting.

[235] What do you find in a geometric junkyard? (A rectangle.)

[236] I have no fault to find with those who teach geometry. That science is the only one which has not produced sects; it is founded on analysis and on synthesis and on the calculus; it does not occupy itself with probable truth; moreover it has the same method in every country.

—Frederick the Great (1712–1786), King of Prussia (1740–1786)

[237]

In the fourth century B.C., according to Proclus, Alexander the Great asked his teacher, Menaechmus, [1] for a short cut to geometry and received the reply "Oh, King, for traveling over the country, there are royal roads for kings, but in geometry there is one road for all."

—as quoted in *Memorabilia Mathematica*,
by Robert Edouard Moritz, MAA,
Spectrum Series, reprint from 1914, quote #901.

[238]   "We do not listen with the best regard to the verses of a man who is only a poet, nor to his problems if he is only an algebraist; but if a man is at once acquainted with the geometric foundation of things and with their festal splendor, his poetry is exact and his arithmetic musical."

—Ralph Waldo Emerson (1803–1882)
Society and Solitude, 1870, ch. 7, *Work and Days*

---

[1]Menaechmus is the inventor of the conics. Platonists would reject the term "inventor," preferring "discoverer" instead. (See item **[673]** for a definition of Platonism.)

[239] It has been observed that the ancient geometers made use of a kind of analysis, which they employed in the solution of problems, although they begrudged to posterity the knowledge of it.

—René Descartes (1596-1650) from his *Regulae, Rule IV,*
as quoted in *Memorabilia Mathematica,* by
Robert Edouard Moritz, MAA, Spectrum Series,
reprint from 1914, quote #1874.

Note: See also item **[580]** "Is Progress Monotonically Increasing?" concerning progress and its presumed inevitability.

[240] It is astonishing that this subject [projective geometry] should be so generally ignored, for mathematics offers nothing more attractive. It possesses the concreteness of the ancient geometry without the tedious particularity, and the power of analytical geometry without the reckoning, and by the beauty of its ideas and methods, illustrates the esthetic generality which is the charm of higher mathematics, but which the elementary mathematics generally lacks.

—From a report on secondary mathematics, 1894,
as quoted in *Memorabilia Mathematica,* by
Robert Edouard Moritz, MAA, Spectrum Series,
reprint from 1914, quote #1876.

[241]

 In contrast to V. I. Arnold, in item [12] who decries the abandonment of geometry in the classroom by "zealots" who prefer abstraction, Voltaire sees nothing in geometry but folly. Here is how he puts it.

—JdP

... but of all the sciences, the most absurd, and that which in my opinion, is most calculated to stifle genius of every kind, is geometry. The objects about which this ridiculous science is conversant, are surfaces, lines, and points, that have no existence in nature.

... the geometrician makes a hundred thousand curved lines pass between a circle and a right line that touches it, when, in reality, there is not room for a straw to pass there.

Geometry, if we consider it in its true light, is mere jest, and nothing more.

—Voltaire (1694–1778) *The Best Known Works of Voltaire*, Blue Ribbon Books, New York, 1927, pp. 287–288.

GILBERT AND SULLIVAN PARODIES

[242]   Now, if I may digress momentarily from the mainstream of this evening's symposium, I'd like to sing a song that is completely pointless, but is something that I picked up during my career as a scientist. This may prove useful to some of you some day, perhaps, in a somewhat bizarre set of circumstances. It's simply the names of the chemical elements set to the tune: "The Major-General's Song," by Sir Arthur Sullivan, from Gilbert & Sullivan's *The Pirates Of Penzance.*

> There's antimony, arsenic, aluminum, selenium,
> And hydrogen and oxygen and nitrogen and rhenium,
> And nickel, neodymium, neptunium, germanium,
> And iron, americium, ruthenium, uranium,
> Europium, zirconium, lutetium, vanadium,
> And lanthanum and osmium and astatine and radium,
> And gold and protactinium and indium and gallium, (gasp)
> And iodine and thorium and thulium and thallium.

There's yttrium, ytterbium, actinium, rubidium,
And boron, gadolinium, niobium, iridium,
And strontium and silicon and silver and samarium,
And bismuth, bromine, lithium, beryllium, and barium.

Don't you find that interesting? I knew you would. I hope you're all taking notes, because there's going to be a short quiz next period. . .

There's holmium and helium and hafnium and erbium,
And phosphorus and francium and fluorine and terbium,
And manganese and mercury, molybdenum, magnesium,
Dysprosium and scandium and cerium and cesium.
And lead, praseodymium and platinum, plutonium,
Palladium, promethium, potassium, polonium,
And tantalum, technetium, titanium, tellurium, (gasp)
And cadmium and calcium and chromium and curium.

There's sulfur, californium and fermium, berkelium,
And also mendelevium, einsteinium, nobelium,
And argon, krypton, neon, radon, xenon, zinc and rhodium,
And chlorine, carbon, cobalt, copper, tungsten, tin and sodium.

These are the only ones of which the news has come to Hahvard,
And there may be many others but they haven't been discahvered.

And now, may I have the next slide please?

—Tom Lehrer

[243] I Am the Very Model of a Doctor of Philosophy

Music: *I Am the Very Model of a Modern Major General* from *The Pirates of Penzance* (1879), by Gilbert and Sullivan

I am the very model of a doctor of philosophy,
I argue trivialities with elegant verbosity.
I formulate new doctrines that can set my students' heads a-spin
and count how many angels waltz and tango on the heads of pins.

In councils and committee rooms I never am denied my say,
I know that proper arguments outweigh conclusions any day.
And when it's time to casually cite obscure authorities
I have no peer and all can hear my vast superiority.

From atrozine to zygotype I flaunt a great vocabul'ry,
My verbal jousts come charging faster than the US cavalry.
In short for unexceeded intellectual pomposity,
I am the very model of a doctor of philosophy.

I am the very model of a doctor of philosophy,
humility would be for me an act of sheer hypocracy.
And like the writing on the wall (you'll find it in Omar Khayyam),
it cannot be denied at all how wonderfully smart I am.

With glorious detail hear me regale my colleagues endlessly,
about the life and habits of a slug unique to Thessaly,
but when it comes to keeping that appointment I made yesterday...
Did I forget? Sincere regrets. Let's try again next Saturday.

A wondrous store of arcane lore is stashed within my cranium.
I have as much good common sense as grandma's prize geranium.

I HAVE AS MUCH GOOD COMMON SENSE AS GRANDMA'S PRIZE GERANIUM.

In short for unexceeded intellectual pomposity,
I am the very model of a doctor of philosophy.

—Dan Kalman*, American University

GOOD MATHEMATICS

[244] "That's a p-baked idea where p ranges from 0 to 1."
I got this quote of I. J. Good from John Riordan, my emeritus colleague at the Rockefeller University. Before coming to Rockefeller, John
had worked for years at the original Bell Labs building in lower Manhat-
tan. One of his colleagues there was I. J. Good who was, apparently, quite
a character. Good would often speculate about (measure, calculate?) the
p-bakedness of current ideas. He even talked about starting a Journal of
Partially Baked Ideas.
John gave me a second quote about the ebullient I. J. Good. This one
was from somebody at Bell Labs who observed, "It's an ill Good that blows
no wind."

—Larry Harper*, University of California, Riverside

[245] If you disregard the very simplest cases, there is in all of mathematics not a single infinite series whose sum has been rigorously determined.
In other words, the most important parts of mathematics stand without a
foundation.

—Niels H. Abel (1802–1829) In G. F. Simmons, *Calculus Gems*,
New York: McGraw-Hill, Inc., 1992, pg. 188.

H

HAIR

[246]   Descartes had bangs,
a shoulder-length hair-do-
looked like a pirate
or swash-buckling dare-do.

Fermat's hair longish
and parted in the middle-
a quizzical expression
(Fermat's Last Riddle?)

Leibniz wore a wig
in a formal sort of way
like a judge in a courtroom
or an actor in a play.

Lagrange-patrician nose,
a medal on a fob-
his white hair cut
in a page-boy bob.

Fourier had ringlets
nature's own style-
on his cherubic face
a captivating smile.

Gauss favored side-burns
a hat with no brim-
finely chiseled features
distinguished him.

Lobachevsky-clean-shaven
with trimmed black hair-
medals on his chest
and aloofness in his air.

Riemann -granny glasses
and a thick black beard
learned and sincere
was how Reimann appeared.

Cantor-a goatee
and mustache foliation-
a most impressive figure
in any aggregation.

Einstein looked amused
at some abstruse reflection-
his hair shooting off
in every direction.

—Katherine O'Brien, *Mathematics Magazine*,
vol. 47, no. 3, pg. 149, 1974.

[247]   When fortune comes, seize her by the forelock, for, I tell you, she is bald at the back.

—Leonardo da Vinci (1452–1519)

HILBERT

Note: See also item [324] for Kurt Gödel's effect on David Hilbert's research.

[248]   Hilbert in his researches on integral equations, considered infinite sequences that were square summable. Eventually such sequences came to be regarded as points in an infinite dimensional space, and an appropriate "geometry" was developed. Although Hilbert did not take this point of view, such spaces and analogous space of square integrable functions became known as "Hilbert Spaces."

One day Hilbert was attending a mathematical meeting with his colleague Courant. At this meeting it seemed that every other paper referred to this or that Hilbert Space, or this or that property of Hilbert Space. After one of these papers, Hilbert is reported to have turned to Courant and said "Richard, exactly what is a Hilbert Space?" (See Index for other Hilbert quotes.)

HUMILITY

Note: See also the comments of Huston Smith, item [98], for perceived arrogance of scientists.

[249]   Today, we know we are formed from the same stuff as stars. We are a piece—not separate—from this grand, entangled cosmos. And we control—at least to some extent—our destinies.

More impressive still, our brains have learned to spin space-time into black holes and to tame quantum unruliness to operate computers. On good days, we even have the mind to appreciate the grandeur of it all—setting it to music, or art, or equations.

Along the way, of course, we've also learned enough to allow us to destroy much of it without the help of any gods at all. This, in itself, is an awesome sort of power.

So it's not entirely clear whether the discoveries of science have really made us more humble. But they have certainly, as the late physicist Frank Oppenheimer put it, "changed the nature of our humility."

—K.C. Cole, science writer,
Los Angeles Times, 15 Feb. 2001.

[250]   In questions of science, the authority of a thousand is not worth the humble reasoning of a single individual.

—Galileo Galilei (1564–1642) as quoted in *Memorabilia Mathematica*, by Robert Edouard Moritz, MAA, Spectrum Series, reprint from 1914, quote #1528.

[251]   [While we are awake, we] know that we do not dream, and, however impossible it is for us to prove it by reason, this inability demonstrates only the weakness of our reason, but not, as they affirm, the uncertainty of all our knowledge.

. . . And it is as useless and absurd for reason to demand from the heart proofs of her first principles, before admitting them, as it would be for the heart to demand from reason an intuition of all demonstrated propositions before accepting them.

This inability ought, then, to serve only to humble reason, which would judge all, but not to impugn our certainty, as if only reason were capable of instructing us.

Would to God, on the contrary, that we had never need of [reason] and that we knew everything by instinct and intuition! But nature has refused us this boon. On the contrary, she has given us but very little knowledge of this kind; and all the rest can be acquired only by reasoning.

—Blaise Pascal (1623–1662) *Penseés* #282

I–J

I'S AND J'S

[252]

> I'm like a rat within a maze
> When faced with sigma's i's and j's,
> And problems soon become enigmas
> When wrapped in i's and j's and sigma's.
>   —Kenneth E. Boulding, *Views on General System Theory*,
>   edited by H.D. Mesarovic, John Wiley & Sons, Inc., 1964.

[253]

> The i-th power of i seems mysterious.
> But if you take it quite serious,
>   You can show that the following is true:
>   It's e to the minus π over 2.
> You have to admit that's quite curious.
>   —Nick J. Rose

INFINITY

Note: See also "Transcendental numbers" item **[740]**.

[254]

*The mysticism (but not the mystery) of infinity was
eliminated when Georg Cantor (1845–1918) rigorously*

and simply defined it in terms of one-to-one maps between
pairs of sets, *namely, between a set S and proper subsets
of S. (A* **proper subset** *is produced from set S when at least
one element is eliminated.) An infinite population (infinite
cardinality) of a set S is defined intuitively as follows:*

Definition of an infinite set

A set S is **infinite** *(has an* **infinite population***) if you can
find a one-to-one map between S and a proper subset of S.
If no such one-to-one map can ever be found, then set S is
called* **finite** *(has a* **finite population***).*

—JdP

[255]

"Infinity bottles of beer on the wall, infinity bottles of beer, take one
down, pass it around, . . . infinity bottles of beer on the wall." Start that one
up when you aren't sure how long the bus ride will be.

—Lew Lefton, Georgia Institute of Technology

[256] Our minds are finite, and yet even in these circumstances of finitude
we are surrounded by possibilities that are infinite, and the purpose of life
is to grasp as much as we can out of that infinitude.

—Alfred North Whitehead (1861–1947)
Dialogues of Alfred North Whitehead, Hardcover Reprint edition,
Greenwood Publishing Group, February 1977.

[257] The infinite! No other question has ever moved so profoundly the
spirit of man.

—David Hilbert (1862–1943)

[258]

O God! I could be bounded in a nutshell,
and count myself king of infinite space,
were it not that I have bad dreams.

—William Shakespeare (1564–1616) *Hamlet*,
Act 2, Scene 2, lines 254–256.

[259]   We have adroitly defined the infinite in arithmetic by a love knot, in this manner ∞; but we possess not therefore the clearer notion of it.

—Voltaire (1694–1778)

[260]

Teacher: What is infinity?

Student: Infinity is that place which, when you reach it, is still farther ahead.

[261]   The world of learning is so broad, and the human soul is so limited in power! We reach forth and strain every nerve, but we seize only a bit of the curtain that hides the infinite from us.

—Maria Mitchell, (1818–1889), astronomer, first woman elected to the American Academy of Arts and Sciences (1848).
In *Maria Mitchell, Life, Letters, and Journals*, 1896.

[262]

So, naturalists observe, a flea
Hath smaller fleas that on him prey.
And these have smaller still to bite 'em,
And so proceed ad infinitum.

—Jonathan Swift (1667–1745)
Rhapsody of Poetry, written in the interval 1730–1735.

[263]   The solution of the difficulties which formerly surrounded the mathematically infinite, is probably the greatest achievement of which our age has to boast.

—Bertrand Russell (1872–1970)

[264]

Infinity? Wait a minute! That's a car!

—Jay Leno* Entertainer, automobile/motorcycle enthusiast
Response to a list of ordinary words that become specialized in a
mathematical context (*e.g.*, group, prime, field, ring, normal,
infinity), backstage at the *Tonight Show*, 2 Feb. 2001.

INNOCENCE

[265]   Touch a scientist and you touch a child.

—Ray Bradbury, *Los Angeles Times*, Aug. 9, 1976.

[266] Remembrance of Things Past—computers as ultimate innocents.
The biological imperfection of man, at the present state of his evolution, has been discussed at great length. Our critics unfairly neglect the other side of the equation: the complexity of the problems we face. Man is the

only intelligent creature on this planet; as soon as his rational functions evolved to the point that culture was possible, he was bound to explore and inhabit the biosphere to the very margins of possibility, and these margins are neither always comfortable nor assuredly durable. Man is likely to press against similar margins whatever improvements he evolves or designs.

I can, however, think of one function we lack by contrast to electronic computers, namely INNOCENCE. No complex program ever works as an immediate outcome of its prior design. However, computer programs can be perfected ("debugged") by repetitive trial and error—the disasters being subject to erasure so that the system memory is restored to a standard condition. This stabilizes the problem so that it can often be solved. The human condition suffers from the fact that our sins and our guilt are cumulative; we may assuage, we may forgive, but have no way to forget.

—Joshua Lederberg, Nobel Laureate, *Physiology & Medicine 1958*.
The Sciences [NY Academy of Sciences] 9 (12):23–24, 1969.

[267]   If innocence is that quality of not knowing the full realities, then the scientist and the mathematician are falling ever deeper into states of innocence.

How is this so? Once research answers one question, it raises several more. Each advance expands the dimensions of the unknown—we understand the full reality even less than before—we become more and more innocent.

That is a major component of research—a quest for innocence.

—John de Pillis

Note: See also, George Bernard Shaw, item **[688]** on the increasing number of unsolved problems in science.

INQUIRY AND RESEARCH

Integers and Dimension: Beyond Our Reach?

[268]   The trouble with the integers is that we have examined only the small ones. Maybe all the exciting stuff happens at really big numbers, ones we can't get our hands on or even begin to think about in any very definite way. So maybe all the action is really inaccessible and we're just fiddling around. Our brains have evolved to get us out of the rain, find where the berries are, and keep us from getting killed. Our brains did not evolve to help us grasp really large numbers or to look a things in a hundred thousand dimensions.

—Ron Graham, University of California, San Diego
from preface of *Wonders of Numbers: Adventures in Mathematics,
Mind and Meaning* by Clifford Pickover, Oxford Univ Pr, 2000

Note: On whether the human mind is too puny a vessel, see also Charles Darwin, item **[212]**.

INTEGERS AND FRACTIONS

[269] And do not swear by your head, for you cannot make even one hair white or black. Simply let your 'Yes' be 'Yes,' and your 'No,' 'No'; anything beyond this comes from the evil one.

—Holy Bible (New Int'l Ver.) Matthew 5:36–37
An early reference to binary code

Note: For other toggling, either-or references, see Abraham Lincoln, items **[274]**, **[275]**; Eric Schechter, item **[340]**; Martin Gardner, item **[207]**; and the Monty Hall problem, item **[300]**.

The Importance of Understanding Fractions

"One-tenth?! We give only one-tenth for the Lord's work? Surely that is not enough. I say we give one-twentieth!"

[270] *(Best when recited aloud.)* A young man did not want to serve in the army so he decided to deliberately fail the hearing test. "Repeat the following integers after me," the physician instructed. The physician whispered, "Thirty-six." The young man responded, "six!" "Hmm," the physician mused. Then, a little bit louder, the physician said, "Seventy three." "Three!" the young man answered. Speaking louder still, the physician said, "Forty." The young man replied, "Zero!"

—As told by Michael Neumann, University of Connecticut

[271]

28. A perfect number.
In math that is, if you remember:
Its factors sum up to itself
as your makeup makes you yourself.

See, $1 + 2 + 4 + 7$
(that makes 14, not 11)
then add 14 and do the math,
28 is what you'll have!

—Naomi Chesler, University of Wisconsin

[272]   Six is a number perfect in itself, and not because God created the world in six days; rather the contrary is true. God created the world in six days because this number is perfect, and it would remain perfect, even if the work of the six days did not exist.

—St. Augustine (354–430) *The City of God.*

Note: See also item **[162]** for a discussion of cause vs. effect, induction vs. deduction and common confusions between them.

[273]   Grace Hopper enjoyed telling a story about her husband's complaint when they were dating. She quoted him: "When you date a mathematician, you plan a great romantic evening. She doesn't respond. After two hours, she says, 'I've got it! It's zero!' "

—Nancy Whitelaw, *Grace Hopper, Programming Pioneer*,
Scientific American Books for Young Readers, New York,
W.H. Freeman and Company, pg. 21. 1995.

"Psst! How about a nice number on special?"

Note: For more on Grace Hopper, see item **[712]**.

Toggling

[274]   If this is coffee, then please—bring me some tea. But if this is tea, please bring me some coffee.

—Abraham Lincoln (1809–1865),
An early reference to toggling, as quoted in *1, 911 Best Things Anybody Ever Said*, ed. Robert Byrne, 1988.

[275]   If I were two-faced, would I be wearing this one?

—Abraham Lincoln (1809–1865),
Another early reference to toggling. (For another either/or choice, see Eric Schechter, item **[340]**.)

[276]

As long as we are crediting Abraham Lincoln with an early appreciation of toggling, we note some other points of interest:

- *Lincoln's formal education lasted only about one year but he received honorary degrees from Knox College (1860), Columbia (1861), and Princeton (1864).*

- *Lincoln is the only U.S. President to own a patent. Patent: #6469 was granted May 22, 1849 for a device to lift boats over shoals.*

- *There is no Presidential Library for Lincoln. In the sixty years since presidential libraries were authorized by Congress in 1939, starting with the one for Herbert Hoover, about a dozen have been built. However, there is a major collection of Lincoln legal papers in Springfield, Il., for which see http://www.lincolnlegalpapers.org/.*

—JdP

INTEGRALS

[277] Common integration is only the memory of differentiation.

—Augustus De Morgan (1806–1871)

[278] Does anyone believe that the difference between the Lebesgue and Riemann integrals can have physical significance, and that whether say, an airplane would or would not fly could depend on this difference? If such were claimed, I should not care to fly in that plane.

—Richard W. Hamming (1915–1998)
paraphrased from *Amer. Math. Monthly*, 105 (1998), 640–50.

[279] There are in this world optimists who feel that any symbol that starts off with an integral sign must necessarily denote something that will have every property that they should like an integral to possess. This of course is quite annoying to us rigorous mathematicians; what is even more annoying is that by doing so, they often come up with the right answer.

—E. J. McShane *Bulletin of the American Mathematical Society*,
vol. 69, pg. 611, 1963

Note: See also Intuition: Examples that Challenge, items **[283]**–**[305]**.

[280] But just as much as it is easy to find the differential of a given quantity, so it is difficult to find the integral of a given differential. Moreover, sometimes we cannot say with certainty whether the integral of a given quantity can be found or not...

—Johann Bernoulli (1667–1748)

[281]   It seems to be expected of every pilgrim up the slopes of the mathematical Parnassus, that he will at some point or other of his journey sit down and invent a definite integral or two towards the increase of the common stock.

—James Joseph Sylvester (1814–1897)

INTUITION: EXAMPLES THAT CHALLENGE

Note: See also item **[229]** for an example of a deceptively drawn diagram.

[282]   Many things are not accessible to intuition at all, the value of

$$\int_0^\infty e^{-x^2} \, dx$$

for instance.

—J. E. Littlewood (1885-1977)

Note: See Lord Kelvin's remark, item **[352]**, about mathematicians who are strangely comfortable with the same integral above. For comments on the intuition-defying Banach-Tarski paradox, see Wikström, item **[517]**.

[283] Algebra: Unique Solution with Unstated Velocities

Problem

Swimmers Toni and Tuna, at opposite ends of a pool, start a race at the same time. While each swimmer maintains her own constant speed, they first meet at distance a from one end of the pool. After swimming the full length of the pool and reversing direction, they again meet at distance b from the other end of the pool.

Question: How long is the pool?

Answer: The pool length is

[284] $$L = 3a - b.$$

(The swimmers' speeds, although fixed and unique, need not be given in advance.)

[285] Beyond intuition: Formula **[284]** above reveals more than just the pool length L. The following examples show that algebra takes us to places that are well beyond our intuition.

When does a solution L exist? Since the pool length $L > 0$, formula **[284]** above 1, $L = 3a - b > 0$, implies that the

$$\text{Ratio } b/a \text{ is less than } 3,$$

a fact that does not intuitively jump to mind.

   If we set the ratio b/a equal to a new parameter, λ, then necessarily, $3 > \lambda > 0$. To see this, note that

[286] $L = 3a - b > 0$ if and only if $\boxed{3a > b.}$

                    if and only if $\boxed{3 > b/a}$

                    if and only if $\boxed{3 > b/a = \lambda,}$ which implies
                                                $3 > \lambda > 0.$

[287] Swimmers' speeds need not be stated: Formulas **[284]** and **[286]** do not seem to depend on the speeds of the two swimmers, Toni and Tuna. Only knowledge of the two meeting point distances, a and b, is needed to find pool length L.

[288] Yet (behind the scenes), swimmers' speeds are strictly constrained: Let v_1 and v_2 represent the constant speeds of the two swimmers. Then, using the ratio λ from **[286]**, derivation of Formula **[284]** leads to the algebraic relation

[289] $v_2/v_1 = 2 - \lambda$, which implies $\boxed{2 > \lambda > 0}$

To test Displays **[284]**, **[286]**, and **[289]**, we consider the (non-intuitive) consequences in the special case when $\lambda = 1$.

From **[289]**, $2 > \lambda > 0$, so $\lambda = 1$ is a legal "test" value. Here are three consequences of setting $\lambda = 1$:

- [from **[289]**] When $\lambda = 1$, $v_2/v_1 = 1$. Thus, the swimmers are moving with identical speeds. (**Note:** the *ratio* of the swimmers' speeds is important, not the individual magnitudes which, for all we know, could be 90% the speed of light.)

- [from **[286]**] When $\lambda = 1$, we have $a = b$ so the two meeting points are equidistant from each side of the pool.

- [from **[284]**] When $a = b$, the length of the pool $L = 2a$, so the two meeting points, in fact, coincide at the very mid-point of the pool.

[290] **Question:** What is the solution to the swimmers' problem with the following modification?

Only one swimmer reaches the other side
of the pool and reverses direction.

This says one of the swimmers is moving too slowly to reach the other side. The swimmers meet first at distance a from one side of the pool (each going in opposite directions) and then they meet a distance b from the other side of the pool (both going in the same direction).

Algebra: The Potato Paradox

[291]

   (A) You buy 100 pounds of potatoes which are 99% water.

   (B) Overnight drying leaves the potatoes at 98% water.

Question: How much do the potatoes weigh now?

Answer: The dried potatoes weigh 50 pounds. Why? From (A) we have that solid $S =$ is 1% of 100 pounds or $S = 1$ pound. Now (B) tells us that the same solid ($S = 1$ pound) survives as 2% of potato weight P. English-to-algebra says $S = 1.00$ lb. $= 0.02*P$ lb. so P lb. $= 1.00/0.02$ lb. $= 50$ lb.

[292]

For the purposes of this problem, the potato may be replaced by any other fruit or vegetable that is mostly water. But if you were to ask the tobacco hornworm larva, there is no substitute for the potato. Read on.

—JdP

Larva Gets Hooked on the Taste of Potatoes.   The potato is the Achilles' heel of the tobacco hornworm larva, biologist Marta del Campo of Cornell University and her colleagues said. Normally, the hornworm larva will eat virtually any type of foliage. But once it has tasted potatoes, nothing else will do. The larva would rather starve to death than eat anything else, the team reported in [the May 2001 issue of *Nature*]. The key seems to be a chemical called indioside D that is present in potatoes. Indioside D apparently modifies taste buds in the larva's mouth so that other foods taste bad. The discovery may lead to a new way to control the pest.

—Thomas H. Maugh II from his article in the *Los Angeles Times*,
May 14, 2001

Calculus: The Celestial Horn with Infinite Area, Finite Volume

[293]   A quart of ink, which is composed of only finitely many molecules, can paint a surface which is infinite!

To be mathematically specific, here is how a finite volume of ink can fill a funnel to the top—and in so doing, paint the inside, infinite surface.

Problem Form the (horizontal) funnel or Celestial Horn by revolving the function $f(x) = 1/x$, $x \in [1, \infty)$, about the x-axis.

Standard integration techniques show that the volume of the funnel is finite, while the surface area is infinite.

[294]

Question: *How can it be that a **finite volume** of ink covers an **infinite surface**? Of course, this paradox results from our careless assumption that the Celestial Horn is being filled with finitely many physical molecules, each of whose*

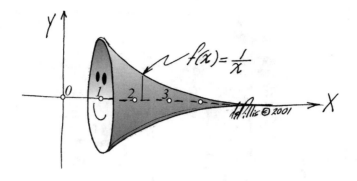

size is fixed and bounded. The finitely many molecules in a quart of ink (a physical reality) are but an approximation model of the finite volume of the Celestial Horn (a mathematial concept).

[295]

Question: *For which values* α *does* $f_\alpha(x) = 1/x^\alpha$ *produce a Celestial Horn with finite volume and infinite area?*

Note: *The term, "Celestial Horn," is due to* **Russ Merris**.

—*JdP*

Calculus: Fable of the Good Queen's Napkin Rings

[296]   A Good Queen gathered all the artisans of her land to fashion golden napkin rings for her royal table. She distributed solid gold spheres of varying sizes to each artisan, noting the following conditions:

- Each napkin ring would be the remainder of the sphere *after* the artisan removes a cylinder from its center. (The removed cylinder is symmetric about the sphere's center.) Of course, this "wasted" cylindrical core of the sphere, once removed, must be returned to the Good Queen.

- The height of each napkin ring is to be the same for all the artisans, namely h units.

- Payment to each artisan is to be made according to the weight of each napkin ring.

The Queen's Golden Napkin Rings

Finally, when all the napkin rings had been crafted, the Good Queen gathered the artisans together and told them they would all be paid the same amount!

"What! We get equal payment for napkin rings made from spheres of different sizes?" the artisans wondered out loud. The Good Queen immediately interrupted the rising din of dissatisfaction, saying, "Oh artisans of the land! Hear me well! I would not be unfair, unjust or unqueenly! Know you, that before you ever started this task, I had used calculus and discovered this fact—the volumes of your wonderful napkin rings would all be equal!"

With a few strokes on the Royal Blackboard, the Good Queen showed her artisans that she spoke the truth. Indeed, it was as the Good Queen had said. The volumes (hence, the weights) were all equal! "How fortunate we are," said the artisans, "we have a queen who is true, fair, and good. And she is good at calculus, too."

Moral: Those who try to be fair and just are too often misunderstood. Especially if they can do calculus.

Probability: von Mises' Birthday Problem

[297]   In 1939, Richard von Mises (1883–1953)[1] raised the

> **Question:** *How many people must be in a room before the probability that at least two people share a birthday (ignoring the year and ignoring leap days) becomes at least 50 percent?*

The nonintuitive answer is that in a group of 23 people, the probability of a birthday match (or collision) is 0.5073, or just over 50%. This assumes that each of the 365 days is an equally likely birthday.

The following graph, which illustrates this fact, also shows that with a group of 41 people, the probability that at least two people share the same birthday is 0.9032, or just over 90%.

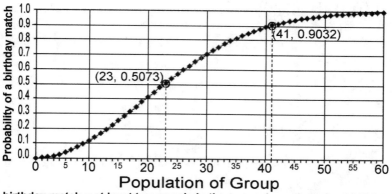

birthday match = at least two people in the group have the same birthday

Probability: Monty Hall Problem

[298]   **The Set-Up:**   You choose one of three inverted cups, one of which conceals a valuable diamond. (Your probability of choosing the cup with the diamond is 1/3.) Now your hostess has one constraint—among the two cups you did not choose, she will always show you an empty cup and then she will discard it. Now, with only two cups in front of you (one of these still has your finger on it) she gives you the option of changing your mind and choosing the other cup.

---

[1]"Über Aufteilungs–und Besetzungs-Wahrscheinlichkeiten." Revue de la Faculté des Sciences de l'Université d'Istanbul, N. S. 4, 145–163, 1939. More accessible might be *Selected Papers of Richard von Mises*, Vol. 2 (Ed. P. Frank, S. Goldstein, M. Kac, W. Prager, G. Szegö, and G. Birkhoff). Providence, RI: *Amer. Math. Soc.*, pp. 313–334, 1964.

The Question: Should you

(a) *always* stay with your original choice, where your probabilty of success is 1/3 or

(b) *always* switch your original choice to the remaining cup?

The Answer: If you always stay put and never switch, your chance of choosing the diamond is 1/3 as it had been from the beginning. But if you *always* switch, the probability jumps to 2/3, thereby doubling your chance of success. The diagram labeled "Monty Hall Problem," (item **[299]**) shows how this happens.

[299]

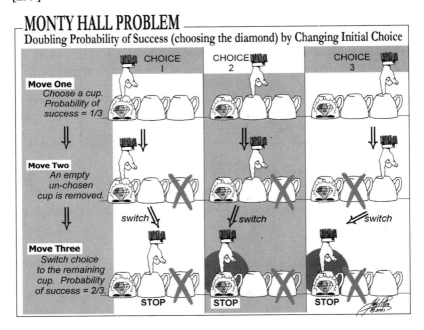

[300] Switching as a Toggling Strategy

If

> "**success**" means we have chosen the cup with the diamond and
>
> "**failure**" means we have chosen an empty cup,

then the diagram above (item **[299]**) shows that by *always* switching after your intial choice, you produce the toggling phenomenon:

- initial **success** (with probability 1/3) toggles to **failure**, and

- initial **failure** (with probability 2/3) toggles to **success**.

Note: For more on toggling in the sense of an either/or choice, see Abraham Lincoln, items [274], [275]; Eric Schechter, item [340]; Martin Gardner, item [207]; and the Bible, item [269].

[301] Large Numbers are Better Than Small Ones

*One way to make the problem easier to understand is to use large numbers instead of the **small** ones we have been dealing with. Replacing large numbers by small ones in a problem usually helps our intuition. (For example, in item [268], Ron Graham describes our inate inability to deal with very large integers.) The opposite is true in this case—large numbers aid our intuition! The details are in item [302] following.*

—JdP

[302] Monty Hall Problem (Large Number Version): You choose one of 1, 000 inverted cups, one of which conceals a valuable diamond. (Your probability of choosing the diamond is only 1/1,000.) Now among the 999 remaining cups, (one of which, most likely, covers the diamond), your hostess discards 998 empty ones. Since you most likely chose an empty cup, the 999th cup your hostess leaves, most likely, is covering the diamond. Now that you have two cups in front of you, she gives you the option of changing your mind and choosing the other cup.

Question: Should you

(a) *always* stay with your original choice, where your probabilty of success is 1/1,000 or

(b) *always* switch and choose the other remaining cup?

Answer: As before, if you always switch, your initial FAILURE (probability 999/1000) toggles to a SUCCESS, and SUCCESS (probability 1/1000) toggles to a FAILURE. (See item [300].) Since there are a large number of cups, your initial choice will almost always miss the diamond.

This is a FAILURE that has probability 999/1000 and is converted to (toggles to) a SUCCESS.

In sum, the only way you can lose, using the "always-switch" strategy, is if you choose the diamond with your first choice, an event with probability 1/1000. That is,

- If your first choice is the cup with the diamond (1 chance in 1000), then after switching, you lose.

- On the other hand, if your first choice is an empty cup (a very likely 999 chances in 1000), then after switching, you win.

Note: The "Monty Hall Problem" diagram, item **[299]**, can also be extended to this case of 1, 000 cups.

[303]

Historical Background Note: In the TV game show of the 1970's, Let's Make a Deal, hosted by Monty Hall, there were three closed doors, one of which concealed a valuable prize. After a contestant selected one of the doors, Monty Hall would open one of the two unselected doors, showing that it did not hide the prize. This door, called a "Zonk," was then eliminated, leaving only two unknown doors in the game. The contestant was then given the option of staying with the originally chosen door or switching to the other unopened door.

—JdP

[304]

In the following somewhat-coherent letter, Monty Hall himself weighs in to disagree with Steven Selvin who published a solution to the Monty Hall problem. Mr. Hall suggests that Professor Selvin may have "manipulated" the figures.

—JdP

Dear Steve:

Thank you for sending me the problem from *The American Statistician*.

Although I am not a student of statistics problems, I do know that these figures can always be used to one's advantage, if I wished to manipulate same. The big hole in your argument of problems is that once the first box is seen to be empty, the contestant cannot exchange his box. So the problems still remain the same, don't they... one out of three. Oh, and incidentally, after one is seen to be empty, his chances are no longer 50/50 but remain what they were in the first place, one out of three. It just seems to the contestant that, one box having been eliminated, he stands a better chance. Not so. It was always two to one against him. And if you ever get on my show, the rules hold fast for you—no trading boxes after the selection.

Next time let's play on my home grounds. I graduated in chemistry and zoology. You want to know your chances of surviving with our polluted air and water?

Sincerely, Monty

—Monty Hall, May 12, 1975
letter to Assistant Professor of Biostatistics, Steven Selvin,
University of California, Berkeley, objecting to Selvin's
published solution to the "Monty Hall problem."

[305]

Geometry: An Expanding Belt for the Earth's Waistline

Problem: The Earth, a sphere with diameter $D = 7,926$ miles, wears a belt snugly around its equator which has length $29,500$ miles. It is decided that the belt must not lie on the ground. It must be lengthened so that it can be propped up exactly one foot above the surface everywhere on the circumference (see diagram).

Question: How much must the belt be lengthened so that it stands one foot above the Earth's surface everywhere?

Answer: A very small increase in belt length will allow the belt to rise one foot above the entire 29, 500 mile circumference of the spherical Earth. In fact, you need only extend the belt by a distance roughly equal to your own height—assuming you are wearing a very tall stove-pipe hat. Specifically, an extension of only 2π feet will do the trick.

Note: The size of the Earth is irrelevant. For a sphere of arbitrary radius, even a radius of zero, the same 2π foot extension of its equatorial belt will raise the belt exactly one foot over the entire circumference.

ITERATION/RECURSION

"I'M SO PROUD OF YOU, HAROLD. YOU ARE ALMOST RECURSING WITHOUT HAVING TO CURSE IN THE FIRST PLACE!"

"THIS IS THE THIRTIETH TIME SHE'S TURNED
ME DOWN.   BUT I CAN'T DECIDE IF IT'S
AN ITERATIVE OR A RECURSIVE THING."

K

KLEIN BOTTLE

[306]

A mathematician named Klein
Thought the Möbius band was divine.
    Said he, "If you glue
    The edges of two,
You'll get a weird bottle like mine."

                        —Anonymous

[307]

Little Green Man

I live in a house like Klein's (little!) bottle.
Imagine a cylinder. Stretch up and out.
Push in the bottom to link with the throttle.
(You take to R^4 to bring this about.)
    A model ingenious and crafty.
    I find it delightful, but drafty.

My hobby is painting. I'm off on a trip.
Performing a feat acrobatic and thrilling.
Painting a stripe on a Möbius strip,
Juggling my paint can to keep it from spilling
    All over the surface (preventable)
    One-sided and nonorientable.
—Katherine O'Brien, *Two Year College Mathematics Journal*,
                  vol. 12, no. 5, front cover, 1981.

[308]

Three jolly sailors from Blaydon-on-Tyne
They went to sea in a bottle by Klein.
Since the sea was entirely inside the hull
The scenery was exceedingly dull.

—Frederick Winsor, *The Space Child's Mother Goose*,
Purple House Press, 2001.

[309]

Knot a Poem

A knot   and another
knot
may   not   be the
same   knot,   though
the   knot   group of
the   knot   and the
other   knot's
knot   group
differ   not;   BUT
if the   knot   group
of a   knot
is the   knot   group
of the   not
knotted

knot,
the knot is
not
knotted.

—Tim Poston, Chief Scientist, Digital Medicine Lab,
Johns Hopkins, Singapore

[310] (At the University of California, Santa Barbara:) Marty Scharlemann tells the story of a calculus student who came in for help, and after Marty had worked some problems, the student said, "So what kind of math do you like?" Marty said, "Knot theory." The student said, "Yeah, me neither."

Quoted in *The Knot Book: An Elementary Introduction to Mathematical Theory of Knots*, Colin C. Adams, W. H. Freeman & Co., Jan. 2001, pp. 276—278

KNOWLEDGE

[311] Math illiteracy strikes 8 out of 5 people.

[312]

What we know is not much.
What we do not know is immense.

—Pierre-Simon de Laplace (1749–1827) (Allegedly his last words.)
quoted in *Budget of Paradoxes*, by Augustus De Morgan.

[313]   One must learn by doing the thing; for though you think you know it, you have no certainty until you try.

—Sophocles (ca. 496–406 B.C.)

[314]   He is unworthy of the name of man who does not know that the diagonal of a square is incommensurable with its side.

—Plato (ca 429–347 B.C.)

[315]   Knowledge is petrified opinion.

—Tom Lehrer*

[316]   "Why, Sir, that knowledge may in some cases produce unhappiness, I allow. But, upon the whole, knowledge, *per se*, is certainly an object which every man would wish to attain, although, perhaps, he may not take the trouble necessary for attaining it."

—Samuel Johnson (1709–1784)

From *Life of Samuel Johnson, LL.D.*, by James Boswell, paperback reprint edition (August 1979) Viking Press.

L

LAPLACE

[317] I never come across one of Laplace's "Thus it plainly appears" without feeling sure that I have hours of hard work before me to fill up the chasm and find out and show how it plainly appears.

> —N. Bowditch (1773–1836) as quoted in *Teaching and History of Mathematics in the U.S.*, by Cajori, 1896.

Note: See also Obvious, items **[505]**–**[508]**.

[318] A mathematician of the first rank, Laplace quickly revealed himself as only a mediocre administrator; from his first work we saw that we had been deceived. Laplace saw no question from its true point of view; he sought subtleties everywhere; had only doubtful ideas, and finally carried the spirit of the infinitely small into administration.

> —Napoleon I (1769–1821)

LOGIC AND REASON

Note: See item **[377]** for comments from Marston Morse, a mathematician who claims that logic is less important to the development of mathematics than "mysterious powers" and beauty.

[319]   Like Molière's M. Jourdain, who spoke prose all his life without knowing it, mathematicians have been reasoning for at least two millennia without being aware of all the principles underlying what they were doing. The real nature of the tools of their craft has become evident only within recent times. . . .A renaissance of logical studies in modern times begins with the publication in 1847 of George Boole's *The Mathematical Analysis of Logic.*

—Ernest Nagel and James R. Newman
Gödel's Proof, New York University Press, 1986, pg. 39

Note: See also J.L. Heilbron, item **[594]**, for a discussion of the art of deduction learned late.

[320]   Logic is the art of going wrong with confidence.

—Morris Kline (1908–1992)

[321]   You either believe in the law of the excluded middle, or you don't.

—Lew Lefton, Georgia Institute of Technology

[322]   Logic is the hygiene the mathematician practices to keep his ideas healthy and strong.

—Hermann Weyl (1885–1955)
The American Mathematical Monthly, November, 1992.

[323]   Every science that has thriven has thriven upon its own symbols: logic, the only science which is admitted to have made no improvements in century after century, is the only one which has grown no symbols.

—Augustus De Morgan (1806–1871) *Transactions Cambridge Philosophical Society*, vol. X, 1864, pg. 184.

[324]   Built-in Limitation of Reason: Kurt Gödel's Incompleteness Theorem

Kurt Gödel's Incompleteness Theorem, announced in 1931, says that within any "reasonable" logical system, not every true statement can be proved—there will always be true statements that must be left out of any list of all provable statements that the system can possibly produce. Patching the gap by adding a true-but-unproven statement (as a new axiom, say) merely creates another, slightly

larger, "reasonable" system which, once again, leaves out other, true statements, i.e., those that can never be proven within the new, "bigger" system.

Gödel's success brought David Hilbert's research, in one area, to a screeching halt. Hilbert was trying to formalize mathematics and Gödel's fact of unavoidable incompleteness stood in stark contradiction to Hilbert's plan to set down rules and structure of logic that would mechanically lead to all *true or provable statements. Since then, the fact that there will always be true statements that are forever inaccessible has been a source of continuing debate and discussion.*

Which specific *statements are unprovable? Is Goldbach's conjecture one such true but never-to-be-proven statement? (For which see item* **[331]**.*)*

In the following, we hear a discussion at Table Seven of the Starlight Café (See pg. 319) which outlines the main ideas behind Gödel's Incompleteness Theorem.

–JdP

"What do you mean it's impossible to prove everything in mathematics?" Anvil Willie asked Hutch one more time. "My gosh! Mathematicians have been going at it for thousands of years. Shouldn't they be done by now? Have they no work ethic?"

"It's not a question of enough time, Willie," Hutch replied. "It's not even a matter of how clever we are. You see, Kurt Gödel showed in 1931— fairly recently—that within the logic of any "reasonable" mathematical system, like systems with arithmetic which you can program into a computer, there will always be true statements you just cannot prove. Although these statements are true, if you reason with tools within the system, you'll never know it."

Cordelia broke her silence. "I know what you mean!" she exclaimed triumphantly. "I'm more familiar with literature and poetry. But you're saying exactly what Hamlet said to his companion: 'There are more things in heaven and earth, Horatio, than are dreamt of in your philosophy.' There is more 'out there' than we will—or can—ever know."

"Before Cordelia here flies off to the moon or somewhere," Anvil Willie interrupted, "tell me how can you prove that you can't prove something?"

Hutch moved her coffee cup aside and placed her lined pad on the table. "It's quite complicated, but I think I can give you the sense of Gödel's argument, OK?" She looked at each of us in turn—Cordelia, Anvil Willie, and me. We each responded with a nod.

We were committed.

"I'm going to be very imprecise," Hutch said, "but I think we can still get an intuitive idea as to why our reasoning powers will always fail. There will always be some true statements that are unknowable and unreachable."

As she began to write, Hutch outlined this sequence of key ideas:

Step 1. "Reasonable" Systems: "Gödel's Box"

[325] Any "reasonable" system will be characterized by the following three requirements, which we diagram as a three-part "Gödel box":

[326]

TABLE 1. **Three Characteristics of a Reasonable System: "Gödel's Box" of statements and their negatives that can be proved or disproved**

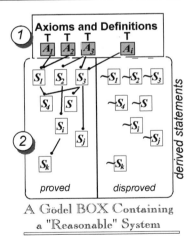

A Gödel BOX Containing a "Reasonable" System

1 **Axioms and Definitions:**
There are initial accepted, or true statements from which all other provable, true statements follow.

2 **Logical Steps:** Starting with these axioms and definitions, certain rules (such as substitution and *modus ponens* which is defined in **[659]**) are used to produce new (proved) statements. As with a digital computer, arithmetic is possible within the system. Each statement S is paired with $\sim S$, its negative, or opposite (disproved) statement.

3 **Consistency:** The axioms and rules cannot derive (prove) both statement S and its negative statement, not S, indicated as $\sim S$. If both S and $\sim S$ are provable, then such an inconsistent system would allow us to prove all statements! The Gödel box splits into two classes: the True (provable) statements and the False (disproved) statements.

[327] Here is an example in which we pass from a primitive statement (definition of even and odd integers) to a deduced property of these integers.

Start with

[328]

TRUE (definition)
An integer is *even* if and only if division by 2 leaves remainder of 0. (definition)
An integer is *odd* if and only if division by 2 leaves a remainder of 1. (definition)

Arrive, by Logical Steps, at

[329]

TRUE (provable)	FALSE
S:	$\sim S$:
Always, the sum of any two odd integers is an even integer.	Sometimes, the sum of two odd integers is an odd integer.

The left statement S is True, and the opposite statement, $\sim S$, on the right-hand side, is False.

Hutch provided the specific logical steps (*i.e.*, a proof). Thus, the definition of item **[328]** above produces a theorem, item **[329]**, whose proof is contained in a Gödel box as illustrated in item **[326]**.

Step 2 **What Is Completeness?**

[330] The Gödel box of item **[326]** diagrams a consistent system which means we cannot prove too much. That is, we cannot show that both S and $\sim S$ are true. On the other end of the scale, we say a system is **complete** if it proves or disproves everything possible—that is, every sentence S the system can produce is provable or disprovable.

Gödel's Incompleteness: What Gödel showed was that any consistent system is incomplete in the sense that many true sentences, created by the symbols in the language, are not provable or derivable within the system. Their truth or falsity must be confirmed OUTSIDE the system. Some statements are wallflowers, simply left out of the logic dance which takes place inside the Gödel box (item **[326]**).

[331] Goldbach Conjecture: One example of a possible wallflower—a TRUE but UNPROVABLE statement *may* be the Goldbach conjecture, which says, simply,

Every even integer greater than two is the sum of two prime integers.

So far as we know, no one has come up with a proof of this statement. Computer searches have shown the Goldbach conjecture is valid for every even integer that has been tested so far. This means that only finitely many integers have been tested and a finite number of successes is not a proof—even if the list of successes reaches from here to the moon.

Step 3 Theorems and Logical Steps Become Integers.

[332] Gödel ingeniously replaced logical statements in normal language by certain integers. More exactly, the symbols used to create sentences were associated with assigned integers. For example, the symbols for "not," "or," "there exists," "if...then," "equals," etc., receive integer assignments. This allows sentences contructed from these symbols to receive integer values. For example, the entire sentence, "There exists an x such that x is the successor of y." can be represented as a single integer.

Step 4 Why the Integer Model Is Important.

[333] If we can represent statements (in true-false pairs) and proofs (*i.e.*, sequences of statements) of a logical system as Gödel integers, then the properties of the statements are mirrored, or transferred, to the realm of integers. Once language is thus "arithmetized," properties of the language are linked to properties of integers. This linkage provides another avenue for understanding.

Step 4a A Defense of Abstraction:

[334] Hutch diverted from the discussion of incompleteness to point out another important consequence of Gödel's integer scheme. When integers can represent natural language (statements) as well as the logical steps (instructions as sequences of statements), then we are describing the von Neumann architecture for computers. John von Neumann, who was very aware of Gödel's work, is generally credited with the idea that both data and instructions for processing those data, could be represented as integers in a digital computer. Merely a coincidence?

Once again, what seems to be abstract and irrelevant on Monday, becomes concrete and essential on Tuesday. Okay, maybe Friday. See the section, "Abstraction," items [7]–[11], which support abstraction along with remarks of V.I. Arnold in item [12], who does not.

Step 5 A Natural Language Example.

[335]   Anvil Willie wanted to know how Gödel showed that some true statements must always be left out of any reasonable system. Hutch obliged with a true statement that is nonprovable—that is, it lies "outside the Gödel box" of all sentences derived from the primitive axioms and definitions. The trick is to construct a sentence that is self-referential.

Hutch provided a self-referential sentence—it described itself in the first person. Taking a bit of license for the sake of intuition, Hutch set down this "Gödel-like" sentence:

[336]

An Unprovable, True Gödel-like Sentence
I am only provable outside the Gödel Box.

[337]   **Case I. If Sentence of item [336] is False:**   (We now argue that if Sentence [336] is False, then it is True, clearly not a good thing.)

If we assume Sentence [336] is **False**, then we must disbelieve it. That is, we believe its opposite which says Sentence [336] is provable only inside the Gödel box. All right then, let us put our Sentence [336] inside the Gödel box. But wait! This forces our (False) sentence to join the other sentences of the Gödel box, all of which are **True**. This contradicts our working assumption I. which says Sentence [336] is False. Assumption I. Doesn't work—it is untenable and unacceptable.

[338]   **Case II. If Sentence (4) is True:**   If Sentence [336] is **True**, then we believe what it says, namely, that it is derivable by axioms and definitions that exist outside the Gödel box. In other words, Case II says the

Gödel box is incomplete. All right, then, let us put our sentence [336] outside the Gödel box where (True) Sentence [336] says it belongs. Indeed, the true sentences inside the Gödel box have no way of creating and proving Sentence [336] (or else it would have been inside the box to begin with!). Our job is done. Case II is valid—the Gödel box is incomplete.

This is the essence of Gödel's Incompleteness Theorem, item [324].

Note: For a nice exposition, see the book, *Gödel's Proof* by Ernest Nagel & James R. Newman, New York University Press, 1986.

[339] Analogy Is Not Proof

After Hutch developed these five steps, the questions began. What does it MEAN if there are true but unprovable statements?

"Sometimes in life, you have to go beyond the system to understand the system," Cordelia volunteered.

Anvil Willie noted that

X-ray machines take us beyond our five senses and produce important information that is otherwise, invisible.

Hutch added that sometimes one has to go outside of one's home dimension, too. For example,

if you are on a one-dimensional line, you don't know if it is curved until you look at yourself in two dimensions. If you are on a two-dimensional sheet, then you won't know if it is dented or bowl-shaped until you go beyond two dimensions to look at yourself in three dimensions.

Once we started pretending we were lines and sheets, we knew we were getting tired. Cordelia said she felt like Sisyphus, rolling ideas up an intellectual hill, only to watch them escape her grasp and to roll back to the bottom.

We looked at each other in silence. Yes, it was definitely time to go home.

—John de Pillis from *Starlight Café Conversations:
An Illustrated Dictionary from Table Seven.* (See pg. 319)

[340]   Even a mathematician must accept some things on faith, or learn to live with uncertainty.

—Eric Schechter, Vanderbilt University
Handbook of Analysis and Its Foundations, Academic Press, 1996,
commenting on the choice forced on us by
Gödel's Incompleteness Theorems.

Note: See also item **[207]** for Martin Gardner's thoughts on uncertainty and choice. For Abraham Lincoln on either/or toggling, see items **[274]**, **[275]**.

[341]   Logic is neither a science nor an art, but a dodge.

—Benjamin Jowett (1817–1893) in James R. Newman (ed.)
The World of Mathematics, New York: Simon and Schuster, 1956.

"FOOD IS <u>NECESSARY</u>, TOO? I THOUGHT
THE PICNIC BASKET WAS <u>SUFFICIENT!</u>"

[342]   If we have a statement such as "3 is greater than 9/12" about the rational number 9/12 and containing a name "9/12" of this rational number, one can replace this name by any other name of the same rational number, for instance, "3/4." If we have a statement such as 3 divides the denominator of "9/12" about a name of a rational number and containing the name of this name, one can replace this name of the name by some other name of the same name, but not in general by the name of some other name, if it is a name of some other name of the same rational number.

—J. Barkley Rosser *Logic for Mathematicians*,
McGraw-Hill Book Co, pg. 52, 1953.

[343]

"...if the ocean was whiskey and I was a duck,...

Looks like the duck is caught in a contrapositive universe again.

'Rye Whiskey' in Contrapositive

Statements

If the ocean was whiskey and I was a duck,
I'd swim to the bottom and never come up.
But the ocean ain't whiskey and I ain't no duck,
So I'll play Jack-O-Diamonds and trust to my luck.
For it's whiskey, Rye whiskey, Rye whiskey I cry,
If I don't get Rye whiskey I surely will die.

If I never reach bottom or sometimes come up,
Then the ocean's not whiskey or I'm not a duck.
But my luck can't be trusted or the cards I'll not buck.
So the ocean is whiskey or I am a duck.
For it's whiskey, Rye whiskey, Rye whiskey I cry.
If my death is uncertain then I get whiskey (Rye).

—W. P. Cooke, *The American Mathematical Monthly*,
vol. 76, no. 9, 1969, pg. 1051.

[344]   The Contrapositive

Note: **Recall the definition of the contrapositive** of implication "*P* implies *Q*" is the implication "~ *Q* implies ~ *P*" (read "not *Q* implies not *P*.") These two implications are equivalent in the sense that they carry the same meaning. See item [30] for details.

[345]   But the fact that some geniuses were laughed at does not imply that all who are laughed at are geniuses. They laughed at Columbus, they laughed at Fulton, they laughed at the Wright brothers. But they also laughed at Bozo the Clown.

—Carl Sagan (1934–1996), from *Broca's Brain:*
Reflections on the Romance of Science, Ballantine Books, 1993.

Note: See also Logic and Reason, from item [319].

[346]   Man is born, not to solve the problem of the UNIVERSE, but to find out where the problem applies, and then to restrain himself within the limits of the comprehensible. His faculties are not sufficient to measure the actions of the universe; and an attempt to explain the outer world by reason is, with his narrow view, vain. The reason of man and the reason of the Deity are two different things.

—Johann Wolfgang von Göthe (1749–1832) quoted
in an article by Walter Orr Roberts: *American Scientist*,
Vol. 55, No. 1 March 1967, pp. 3–14.

Note: See also Kurt Gödel's Incompleteness Theorem, item [324], which says some true statements will never be proved.

M

MATHEMATICIANS ARE AMUSING, DIFFERENT, UNDERAPPRECIATED

[347]   Mathematicians are like Frenchmen: whatever you say to them, they translate it into their own language and forthwith, it is something entirely different.

—Johann Wolfgang von Göthe (1749–1832)
Maxims and Reflexions, 1829.

[348]   Hearing these panelists [of writers and actors] discuss my world [of mathematics] was an eye-opener. . . .They are always searching for the best image for their works. Andrew Wiles at lunch discussing French history is not a powerful image. Andrew Wiles in the attic for seven years is, for them, ideal. Most disconcerting to us, are [their] images that join mathematics and madness, a theme in almost all the popular works.

—Joel Spencer, Courant Institute, N. Y. University
Notices of the Amer. Math. Soc., Opinion section,
Feb. 2001, pg. 165. (Commenting on a seminar with
participants from both the theater and the sciences.)

Note: See also comments of Andrew Wiles, item **[182]**.

[349]

Math never gets into the story, everyone else gets the credit.

—Tony Chan, University of California, Los Angeles
As reported in the *Los Angeles Times*, July 14, 1998, page 1.

Note: See also Stan Osher, item **[137]**, for related comments.

[350]   Biographical history, as taught in our public schools, is still largely a history of boneheads: ridiculous kings and queens, paranoid political leaders, compulsive voyagers, ignorant generals—the flotsam and jetsam of historical currents. The men who radically altered history, the great scientists and mathematicians, are seldom mentioned, if at all.

—Martin Gardner, science writer
The American Mathematical Monthly, Dec. 1994.

[351]

Mini-Profiles

Archimedes stick in hand,
Traced his tombstone in the sand.

KhayyAm laid cubics on the line
But better known for a jug of wine.

Fibonacci couldn't sleep
Counted rabbits instead of sheep.

Fermat found margins a handy place,
But all too soon ran out of space.

Evariste Galois fought a duel
Fate was ruthless, fate was cruel.

Hamilton crossed a Dublin bridge
Carved graffiti on its ridge.
Emmy Noether-Adam's rib
Antedating women's lib.
Gödel-giant stride agility
Decided undecidability.
Bourbaki keeping fit and nifty
Component parts retire at fifty.

> —Katherine O'Brien, *Mathematics Magazine*,
> vol. 48, no. 4, pg. 199, 1975.

[352]   A mathematician is one to whom

$$\int_0^\infty e^{-x^2}\, dx = \frac{\sqrt{\pi}}{2}$$

is as obvious as twice two makes four is to you.

> —William Thomson (later Lord Kelvin) (1824–1907))
> as quoted in *Memorabilia Mathematica*, by Robert Edouard Moritz,
> MAA, Spectrum Series, reprint from 1914,
> quote #822.

Note: See also "Intuition: Examples that Challenge," items **[283]**–**[305]**, where this same integral is given by J.E. Littlewood as a challenge to intuition.

MATHEMATICIANS ARE HIGHLY SKILLED

[353] [Mathematicians] do everything from sharpening fuzzy images for police and astronomers to designing new materials from scratch. Surprisingly, the mathematician accomplishes all these diverse tricks with a comparatively small set of tools—rather like a skilled carpenter who can create an elaborately carved dollhouse or rebuild a garage with hammer, saw, lathe, nails, sandpaper and shellac. Whether the problem is simulating weather or seeing stars, mathematicians always begin by turning things into numbers and relationships into equations. And that requires a clear understanding of the situation at hand.

> —K.C. Cole, science writer
> *Los Angeles Times*, July 14, 1998, page 1.

[354]

"Just another mathematician, I suppose."

The mathematician has reached the highest rung on the ladder of human thought.

—Havelock Ellis (1859–1939) *The Dance of Life*

[355]

The Mathematician

In midair somewhere
he lays an axiomatic floor.
On it he sets a hypothetical plank
on which he raises a logical ladder
which he proceeds to climb.
    There is risk, suspense and drama:
    any loose rung, any misstep fatal.
At the proper confluence of space and time
he steps off onto a higher platform
with a broader panorama.
The whole thing is fabrication.
But so was Creation.

—Katherine O'Brien, *Mathematics Magazine,*
vol. 55, no. 4, pg. 235, 1982.

[356]   Archimedes will be remembered when Aeschylus is forgotten, because languages die and mathematical ideas do not. 'Immortality' may be a

silly word, but probably a mathematician has the best chance of whatever it may mean.

—Godfrey H. Hardy (1877–1947) *A Mathematician's Apology*,
London, Cambridge University Press, 1941.

[357]   A scientist worthy of his name, above all a mathematician, experiences in his work the same impression as an artist; his pleasure is as great and of the same nature.

—Henri Poincaré (1854–1912)

MATHEMATICIANS ARE UNDERSKILLED

[358]   We are in the ordinary position of scientists of having to be content with piecemeal improvements: we can make several things clearer, but we cannot make anything clear.

—Frank P. Ramsey (1903–1930)

Note: Frank Ramsey was a British philosopher who died young but left many important ideas: the improvement of Bertrand Russell's Theory of Types, the distinction between logical and semantic paradoxes, the subjective analysis of probability, the Ramsey-Sentence for eliminating overt reference to theoretical entities, and the Theory of Truth.

—JdP

[359]   I suspect so; but you are speaking, Socrates, of a vast work. What do you mean? I said; the prelude or what? Do you not know that all this is but the prelude to the actual strain which we have to learn? For you surely would not regard the skilled mathematician as a dialectician [i.e., one who discusses and reasons by dialogue]?

Assuredly not, he said; I have hardly ever known a mathematician who was capable of reasoning.

—Plato (ca 429–347 B.C.) *The Republic, Book 7.*

[360]

"OK, Dad, you told me how to use the <u>reasonable doubt</u> argument with Ms. Feeney. But you didn't tell me how to sell her on <u>doubtful reasoning.</u>"

Thinking like a mathematician has its advantages, I suppose. But I have been rejected (again) for jury duty because I have trouble with "reasonable doubt!"

—Agnes M. Kalemaris*, SUNY, Farmingdale

[361]

Riding through some very heavy clouds, three balloonists became hopelessly lost. Finally, through a break in the clouds they saw someone on the ground.

"Where are we?" shouted the balloonists.

"You are up in a balloon," was the reply.

"Are you a mathematician?"

"Yes, but how did you know?"

"Well, what you said was correct but of absolutely no use."

—Anonymous

[362]   In my opinion, a mathematician, insofar as he is a mathematician, need not preoccupy himself with philosophy—an opinion, moreover, which has been expressed by many philosophers.

—Henri Lebesgue (1875–1941) *Scientific American*
vol. 211 (September, 1964) pg. 129.

MATHEMATICS AND ART

[363]   The mathematician's best work is art, a high perfect art, as daring as the most secret dreams of imagination, clear and limpid. Mathematical genius and artistic genius touch one another.

—Gösta Mittag-Leffler (1846–1927)

[364]   Guided only by their feeling for symmetry, simplicity, and generality, and an indefinable sense of the fitness of things, creative mathematicians now, as in the past, are inspired by the art of mathematics rather than by any prospect of ultimate usefulness.

—Eric Temple Bell (1883–1960)

MATHEMATICS AND BEAUTY

[365]   Mathematics, rightly viewed, possesses not only truth, but supreme beauty—a beauty cold and austere, like that of a sculpture, without appeal to any part of our weaker nature, without the gorgeous trappings of painting or music, sublimely pure, and capable of a stern perfection such as only the greatest art can show.

—Bertrand Russell (1872–1970) *The Study of Mathematics:
Philosophical Essays* (London, 1910), pg. 73.

[366]   The bottom line for mathematicians is that the architecture has to be right. In all the mathematics that I did, the essential point was to find

the right architecture. It's like building a bridge. Once the main lines of the structure are right, then the details miraculously fit. The problem is the overall design.

—Freeman Dyson, "Freeman Dyson: Mathematician, Physicist, and Writer." Interview with Donald J. Albers, *The College Mathematics Journal*, vol 25, no. 1, January 1994.

[367] You may object that by speaking of simplicity and beauty I am introducing aesthetic criteria of truth, and I frankly admit that I am strongly attracted by the simplicity and beauty of the mathematical schemes which nature presents us. You must have felt this too: the almost frightening simplicity and wholeness of the relationship, which nature suddenly spreads out before us.

—Werner Karl Heisenberg (1901–1976): communication to Einstein.

Note: See also "Heisenberg Uncertainty Principle," item **[129]**.

[368] The mathematician is fascinated with the marvelous beauty of the forms he constructs, and in their beauty he finds everlasting truth.

—J.B. Shaw, "Mathematics—The Subtle Fine Art," in W.L. Schaaf (ed.), *Mathematics: Our Great Heritage*, New York, Harper, 1948, pg. 50.

[369] As for everything else, so for a mathematical theory: beauty can be perceived but not explained.

—Arthur Cayley (1821–1895) as quoted in *The World of Mathematics*, James R. Newman (ed.) New York: Simon and Schuster, 1956.

[370]
John Brockman: Earlier you mentioned the word beauty. What's with beauty?
Reuben Hirsch: Fortunately, I have an answer to that. My friend, Gian-Carlo Rota, dealt with that issue in his new book, *Indiscrete Thoughts*. He said the desire to say "How beautiful!" is associated with an insight. When something unclear or confusing suddenly fits together, that's beautiful. Maybe there are other situations that you would say are beautiful besides that, but I felt when I read that that he really had something. Because we talk about beauty all the time without being clear what we mean by it;

it's purely subjective. But Rota came very close to it. Order out of confusion; simplicity out of complexity; understanding out of misunderstanding; that's mathematical beauty.

—Reuben Hersh, from an interview with John Brockman,
www.edge.org/documents/archive/edge5.html, Feb 1997.

[371] The mathematician's patterns, like the painter's or the poet's must be beautiful; the ideas, like the colors or the words must fit together in a harmonious way. Beauty is the first test: there is no permanent place in this world for ugly mathematics.

—Godfrey H. Hardy (1877–1947) *A Mathematician's Apology*,
London, Cambridge University Press, 1941.

[372] Mathematics has beauties of its own—a symmetry and proportion in its results, a lack of superfluity, an exact adaptation of means to ends, which is exceedingly remarkable and to be found only in the works of the greatest beauty. When this subject is properly... presented, the mental emotion should be that of enjoyment of beauty, not that of repulsion from the ugly and the unpleasant.

—Jacob William Albert Young (1865–1948), University of Chicago.
In H. Eves *Mathematical Circles Squared*,
Boston: Prindle, Weber and Schmidt, 1972.

[373] We ascribe beauty to that which is simple; which has no superfluous parts; which exactly meets its end; which stands related to all things; which is the mean of many extremes.

—Ralph Waldo Emerson (1803–1882) *The Conduct of Life*

[374] A peculiar beauty reigns in the realm of mathematics, a beauty which resembles not so much the beauty of art as the beauty of nature and which affects the reflective mind, which has acquired an appreciation of it, very much like the latter.

—Ernst Eduard Kummer (1810–1893)

[375] Many arts there are which beautify the mind of man; of all other none do more garnish and beautify it than those arts which are called mathematical.

—Sir Henry Billingsley (1535–1606)

Note: The first translation of Euclid's Elements into English was made by Sir Henry Billingsley, then Mayor of London, and published in 1582 by John Daye. It is said that Billingsley undertook this translation because it would be useful for merchants.

—*JdP*

Note: See also Usefulness/Relevance, items **[415]**–**[433]**.

[376]

> ...Euclid alone
> Has looked on Beauty bare. Fortunate they
> Who though once only and then but far away,
> Have heard her massive sandal set on stone.
> —Edna St. Vincent Millay (1892–1950)
> *Collected Poems* (1923), sonnet 22.

[377] ...discovery in mathematics is not a matter of logic. It is rather the result of mysterious powers which no one understands, and in which unconscious recognition of beauty must play an important part. Out of an infinity of designs a mathematician chooses one pattern for beauty's sake

and pulls it down to earth. The creative scientist lives in a 'wildness of logic,' where reason is the handmaiden and not the master.

—Marston Morse (1892–1977) from the article,
"Princeton & Mathematics: A Notable Record," by Virginia Chaplin,
Princeton Alumni Weekly, May 9, 1958.

Note: On the full article from which item **[377]** is extracted, it is interesting to note the editor's comment: "This is probably the longest article to appear in PAW's 58 years of publication and in the Editor's judgment, one of the very best."

MATHEMATICS AND MUSIC

[378]   Music is the pleasure the human soul experiences from counting without being aware that it is counting.

—Gottfried Wilhelm Leibniz (1646–1716) quoted in
Contemporary Immortals by Archibald Henderson,
Ayer Co Pub, 1930: ch. "Albert Einstein."

[379]   Mathematics and Music, the most sharply contrasted fields of intellectual activity which one can discover, and yet bound together, supporting one another as if they would demonstrate the hidden bond which draws together all activities of our mind, and which also in the revelations of artistic

genius leads us to surmise unconscious expressions of a mysteriously active intelligence.

—Hermann von Helmholtz (1821–1894)
Quoted in R.C. Archibald, "Mathematicians and Music,"
The American Mathematical Monthly, vol. 31, January 1924, pg. 1.

Note: See also item **[99]** for **Hermann von Helmholz's** unintended role in the development of the telephone.

[380]   If I were not a physicist, I would probably be a musician. I often think in music. I live my daydreams in music. I see my life in terms of music....I get most joy in life out of music.

—Albert Einstein (1879–1955) In "What Life Means to Einstein"
An Interview by George Sylvester Viereck, from pg. 17 of
the October 26, 1929 issue of *The Saturday Evening Post*.

Note: See also Grace Hopper's comment on personal fulfillment, item **[711]**.

MATHEMATICS AND NATURE

[381]   I contend that each natural science is a real science insofar as it is mathematics.

—Immanuel Kant (1724–1804)

[382] Why, then, does science work? The answer is that nobody knows. It is a complete mystery—perhaps the complete mystery—why the human mind should be able to understand anything at all about the wider universe.

—Timothy Ferris, from *Coming of Age in the Milky Way*, Doubleday, 1989, pg.385.

[383] To those who do not know mathematics it is difficult to get across a real feeling as to the beauty, the deepest beauty of nature. If you want to learn about nature, to appreciate nature, it is necessary to understand the language that she speaks in.

—Richard P. Feynman (1918–1988)

[384] The most incomprehensible thing about the universe is that it is comprehensible.

—Albert Einstein (1879–1955) "Ideas and Opinions" (under *Principles of Physics*), New York, Crown, 1954, pp.224–227.

Note: See also Eugene Wigner, item [417], for comments on the unreasonable effectiveness of mathematics.

[385] Philosophy is written in this grand book—I mean the universe—which stands continually open to our gaze, but it cannot be understood unless one first learns to comprehend the language and interpret the characters in which it is written. It is written in the language of mathematics, and its characters are triangles, circles, and other geometrical figures, without which it is humanly impossible to understand a single word of it; without these, one is wandering in a dark labyrinth.

—Galileo Galilei (1564–1642) *Il Saggiatore* (1623).

[386] The deep study of nature is the most fruitful source of mathematical discoveries.

—Joseph Fourier (1768–1830)

Note: See comment of V.I. Arnold, item [12], claiming that geometry models reality.

[387] The mathematician, carried along on his flood of symbols, dealing apparently with purely formal truths, may still reach results of endless importance for our description of the physical universe.

—Karl Pearson (1857–1936)

[388]   All the effects of nature are but the mathematical results of a small number of immutable laws.

—Pierre-Simon de Laplace (1749–1827)

[389]

"Let's make it simple, OK? When you learn in
your math class how we're in harmony
with Nature, that means <u>you don't eat us!</u>"

There is no science which teaches the harmonies of nature more clearly than mathematics.

—Paul Carus (1852–1919)

[390]   Mathematics is an interesting intellectual sport but it should not be allowed to stand in the way of obtaining sensible information about physical processes.

—Richard W. Hamming (1915–1998)

[391]

   All the pictures which science now draws of nature and which alone seem capable of according with observational fact are mathematical pictures... From the intrinsic evidence of his creation, the Great Architect of the Universe now begins to appear as a pure mathematician.

—Sir James Hopwood Jeans, *Mysterious Universe*, 1930, ch. 5.

[392]   How can it be that mathematics, being after all a product of human thought independent of experience, is so admirably adapted to the objects of reality?

—Albert Einstein (1879–1955) *Sidelights on Relativity*,
Dover, Oct. 1983 pg. 28.

Note: See also Eugene Wigner, item **[417]**, on the unreasonable effectiveness of mathematics.

MATHEMATICS AND POETRY

[393]   It is an open secret to the few who know it, but a mystery and stumbling block to the many, that Science and Poetry are sisters; in so much that in those branches of scientific inquiry which are most abstract, most formal and most remote from the grasp of ordinary sensible imagination, a higher power of imagination akin to the creative insight of the poet is most needed and most fruitful of lasting work.

—F. Pollock

[394]   We especially need imagination in science. It is not all mathematics, nor all logic, but it is somewhat beauty and poetry.

—Maria Mitchell (1818–1889), astronomer, first woman elected to the American Academy of Arts and Sciences (1848). In *Isaac Asimov's Book of Science and Nature Quotations*, ed. Jason Shulman & Isaac Asimov, 1988.

[395]   (Three Haiku)

Note: See also "Computers/System Error Haiku," item **[149]**.

Fire and Ice
> Strange anomaly:
> the flame of intuition
> frozen in rigor.

Faith and Reason
> Strands of axioms
> intertwining with logic
> in convolution.

Truth and Beauty
> Crucible of proof
> outshining alabaster,
> outlasting marble.

—Katherine O'Brien, *The American Mathematical Monthly*, vol. 88, no. 8, 1981, pg. 626.

[396]   It is as great a mistake to maintain that a high development of the imagination is not essential to progress in mathematical studies as to hold...that science and poetry are antagonistic pursuits.

—F. S. Hoffman

Note: See also *Is Progress Monotonically Increasing?* item **[580]**.

[397]   Poets say science takes away from the beauty of the stars—mere globs of gas atoms...I too can see the stars on a desert night, and feel them. But do I see less or more? The vastness of the heavens stretches my imagination—stuck on this carousel my little eye can catch one-million-year-old-light. A vast pattern—of which I am a part...What is the pattern, or the meaning, or the 'why'? It does not do harm to the mystery to know a little about it. For far more marvelous is the truth than any artists of the past imagined it. Why do the poets of the present not speak of it? What men are poets who can speak of Jupiter if he were a man, but if he is an immense spinning sphere of methane and ammonia must be silent?

—Richard P. Feynman, (1918–1988),
The Feynman Lectures on Physics, I-3-6 n.

[398]   Mathesis and Poetry are...the utterance of the same power of imagination, only that in the one case it is addressed to the head, and in the other, to the heart.

—Thomas Hill as quoted in *Memorabilia Mathematica*,
by Robert Edouard Moritz, MAA, Spectrum Series,
reprint from 1914, quote #1125.

[399]   In science one tries to tell people, in such a way as to be understood by everyone, something that no one ever knew before. But in poetry, it's the exact opposite.

—Paul Adrien Maurice Dirac (1902–1984))
In H. Eves, *Mathematical Circles Adieu*,
Boston: Prindle, Weber and Schmidt, 1977.

[400]   Thus metaphysics and mathematics are, among all the sciences that belong to reason, those in which imagination has the greatest role. I beg pardon of those delicate spirits who are detractors of mathematics for saying this.... The imagination in a mathematician who creates makes no less difference than in a poet who invents.... Of all the great men of antiquity, Archimedes may be the one who most deserves to be placed beside Homer.

—Jean Le Rond d'Alembert (1717–1783)
Discours Préliminaire de L'Encyclopédie,
Tome 1, 1967. pp 47–48.

[401]   ...the spector of a glacial [Isaac] Newton portraying the universe as a machine has furthered the impression that science itself is inherently mechanical and inhuman.... Indifference to the interdependence of science

and the humanities, Newton turned a deaf ear to music, dismissed great works of sculpture as "stone dolls," and viewed poetry as "a kind of ingenious nonsense."

—Timothy Ferris from *Coming of Age in the Milky Way*, Doubleday, 1989, pg. 119, citing source *Newton, Principia*, Cajori-Motte translation, pg. 13.

Note: See also Isaac Newton, items **[487]**–**[503]**.

[402]   The mathematics are usually considered as being the very antipodes of Poesy. Yet Mathesis and Poesy are of the closest kindred, for they are both works of the imagination. Poesy is a creation, a making, a fiction; and the Mathematics have been called, by an admirer of them, the sublimest and most stupendous of fictions.

—Thomas Hill

[403]

For a contrary view on the harmony and sisterhood of Science and Poetry, consider the following comments.

—JdP

Poetry is not the proper antithesis to prose, but to science. Poetry is opposed to science, and prose to metre. The proper amd immediate object of science is the acquirement, or communication, of truth; the proper and immediate object of poetry is the communication of immediate pleasure.

—Samuel Taylor Coleridge (1772–1834)
Definitions of Poetry, 1811.

MATHEMATICS AND TRUTH

Note: See also "Certainty," **[121]**–**[130]** including item **[122]**, for K.C. Cole's observations on reliability of science even though it constantly revises itself. For Ludwig Wittgenstein's denial of absolute truth, see item **[106]** and for his debate with Karl Popper, see item **[173]**. For Henry James' affirmation of absolute truth, see **[687]**.

[404]   The critical mathematician has abandoned the search for truth. He no longer flatters himself that his propositions are or can be known to him or to any other human being to be true; and he contents himself with aiming at the correct, or the consistent.

The distinction is not imminently a kind of truth. He is not absolutely certain but he believes profoundly that it is possible to find various sets of a few propositions, each such that the propositions of each set are compatible, that the propositions of each such set imply other propositions, and that the latter can be deduced from the former with certainty. That is to say, he believes that there are systems of coherent or consistent propositions, and he regards it his business to discover such systems. Any such system is a branch of mathematics.

—C.J. Keyser (1862–1947)

[405]   For since the fabric of the Universe is most perfect and the work of a most wise creator, nothing at all takes place in the Universe in which some rule of maximum or minimum does not appear.

—Leonhard Euler (1707–1783)
Methodus inveniendi lineas curvas, pg. 245, 1744.

[406]   A thousand stories which the ignorant tell, and believe, die away at once when the computist takes them in his gripe [sic].

—Samuel Johnson (1709–1784) From *Life of Samuel Johnson, LL.D.*, by James Boswell, Everyman's Library (Knopf), New York.

[407]   I have often been surprised that Mathematics, the quintessence of Truth, should have found admirers so few and so languid.

—Samuel Taylor Coleridge (1772–1834) *The Complete Poetical Works*, vol I, a mathematical problem.

[408]   When you follow two separate chains of thought, Watson, you will find some point of intersection which should approximate the truth.

—Sir Arthur Conan Doyle (1859–1930) spoken by
Sherlock Holmes in *The Disappearance of Lady Francis Carfax*

[409]   Another thing I got from mathematics has meant more to me than I can say. No man who has not a decently skeptical mind can claim to be civilized. Euclid taught me that without assumptions there is no proof. Therefore, in any argument, examine the assumptions. Then, in the alleged proof, be alert for inexplicit assumptions. Euclid's notorious oversights drove this lesson home. Thanks to him, I am (I hope) immune to all propaganda, including that of mathematics itself.

Mathematical "truth" is no "truer" than any other, and Pilate's question is still meaningless. There are no absolutes, even in mathematics.

—Eric Temple Bell (1883–1960)

Note: For more on absolute truth, see the affirmation of Henry James (item **[687]**) and the denial of Ludwig Wittgenstein (item **[105]**). For related comments concerning Wittgenstein and Karl Popper, see item **[174]**.

[410]   Mathematics is the most exact science, and its conclusions are capable of absolute proof. But this is so only because mathematics does not attempt to draw absolute conclusions. All mathematical truths are relative, conditional.

—Charles P. Steinmetz (1865–1923)
quoted in *Men of Mathematics*, E.T. Bell,
New York, Simon and Schuster, 1937.

[411]   There is no absolute knowledge. And those who claim it, whether they are scientists or dogmatists, open the door to tragedy. All information is imperfect. We have to treat it with humility. That is the human condition.

—Jacob Bronowski (1908–1974), *The Ascent of Man*.
Angus and Robertson: Sydney, 1976, pg. 353.

[412]   Have you noticed that the astronomers and mathematicians are much the most cheerful people of the lot? I suppose that perpetually contemplating things on so vast a scale makes them feel either that it doesn't matter a hoot anyway, or that anything so large and elaborate must have some sense in it somewhere.

—Dorothy L. Sayers, with R. Eustace, *The Documents in the Case*,
New York: Harper and Row, 1930, p 54.

Note: See also Dirk Jan Struik, item **[222]**, on the happiness of mathematicians.

[413]   The most distinct and beautiful statement of any truth must take, at last, the mathematial form.

—David Henry Thoreau (1817–1862) from the "Friday" section of
A Week on the Concord and Merrimack Rivers, 1849.

[414]   Mathematics can never tell us whether any alleged fact is true or false. Mathematics, in short, has no more to do with truth than logic has.

To say something is mathematically proved is tantamount to saying that it cannot possibly be true.

—E. V. Huntington (1874–1952)

MATHEMATICS AND USEFULNESS/RELEVANCE

[**415**]   Even people in science don't know what we [mathematicians] do.

—Stanley Osher, University of California, Los Angeles
As reported in the *Los Angeles Times*, July 14, 1998, pg. 1.

[**416**]

The following quote of G.H. Hardy is one of the most frequently cited among mathematicians. This is not only because his prediction was innacurate, but because of the great irony. Results in general relativity and Hardy's own area of pure (i.e., "irrelevant") number theory have since

*found many applications in the fields of nuclear physics,
coding theory, encryption, and other applied areas. Of
what was to come, Hardy hardly knew.*

—JdP

There is one comforting conclusion which is easy for a real mathematician. Real mathematics has no effects on war. No one has yet discovered any warlike purpose to be served by the theory of numbers or relativity, and it seems very unlikely that anyone will do so for many years.

—Godfrey H. Hardy (1877–1947) *A Mathematician's Apology*,
London, Cambridge University Press, 1941.

Note: As item **[684]** shows, Hardy had a somewhat cynical view of what other people regarded as progress in science.

Unreasonable Effectiveness of Mathematics

[417] The enormous usefulness of mathematics in the natural sciences is something bordering on the mysterious and there is no rational explanation of it.

The miracle of the appropriateness of the language of mathematics for the formulation of the laws of physics is a wonderful gift which we neither understand nor deserve.

—Eugene P. Wigner (1902–1995), *Communications on
Pure and Applied Mathematics* 13 (1960) pg. 2 and pg. 14.

[418] For the things of this world cannot be made known without a knowledge of mathematics.

—Roger Bacon (1220–1292) *Opus Majus*
part 4 *Distinctia Prima* cap 1, 1267.

[419] Pure mathematics is on the whole distinctly more useful than applied.... For what is useful above all, is technique, and mathematical technique is taught mainly through pure mathematics.

—Godfrey H. Hardy (1877–1947)

[420] The lack of real contact between mathematics and biology is either a tragedy, a scandal, or a challenge, it is hard to decide which.

—Gian-Carlo Rota (1932–1999) *Discrete Thought*

Study mathematics! Study mathematics you kept saying!
Any "free module" theorems that can help us out <u>now</u>?

[421]   In geometry as in most sciences, it is very rare that an isolated proposition is of immediate utility. But the theories most powerful in practice are formed of situations which curiosity alone brought to light, and which long remained useless without its being able to divine in what way they should one day cease to be so. In this sense, it may be said, that in real science, no theory, no research, is in effect, useless.

—Voltaire (1694–1778)

[422]   It is a pleasant surprise to him (the pure mathematician) and an added problem if he finds that the arts can use his calculations, or that the senses can verify them, much as if a composer found that sailors could heave better when singing his songs.

—George Santayana (1863–1952) In James R. Newman (ed.) *The World of Mathematics*, New York: Simon and Schuster, 1956.

Note: See also Stan Osher, item **[415]** and Tony Chan, item **[349]**, on the perceived usefulness of mathematics.

[423]   Scientific subjects do not progress necessarily on the lines of direct usefulness. Very many applications of the theories of pure mathematics have come many years, sometimes centuries, after the actual discoveries themselves. The weapons were at hand, but the men were not able to use them.

—A. R. Forsyth (1858–1942)

Note: See also, "Is Progress Monotonically Increasing?" item **[580]**.

[424]   One may say that mathematics talks about things which are of no concern to man. Mathematics has the inhuman quality of starlight, brilliant and sharp, but cold. Thus we are cleverest where knowledge matters least: in mathematics, especially number theory.

—Hermann Weyl (1885–1955)
The American Mathematical Monthly,
vol. 58, no. 8, 1951, pg. 523.

Note: See also item **[416]** for G.H. Hardy's famous (mis)statement on the purity and non-applicability of mathematics.

[425]   What science can there be more noble, more excellent, more useful for men, more admirably high and demonstrative, than mathematics?

—Benjamin Franklin (1706–1790)

[426]   No one really understood music unless he was a scientist, her father had declared, and not just a scientist, either, oh, no, only the real ones, the theoreticians, whose language is mathematics. She had not understood mathematics until he had explained to her that it was the symbolic language of relationships. "And relationships," he had told her, "contained the essential meaning of life."

—Pearl S. Buck (1892–1973) *The Goddess Abides, Pt. I*, 1972.

Bayes' Theorem and Specious Reasoning

[427]

Bayes' Theorem describes probabilities that are affected when additional information reduces the event space (the set of all possible events). For example, if a hotel has 6,000 rooms, and 6 of them have no windows, then your chance of being randomly assigned to a no-window room is 6 in 6,000, or 0.001. (The sample space has 6,000 equally likely rooms.) Suppose we are given additional information, namely, that there are 5 rooms with black-ceilings and 4 of them have no windows. Moreover, we also know that we are assigned to one of the 5 black-ceiling rooms (our sample space is thus reduced to 5 equally likely rooms). Now with this additional information, our chance of being randomly assigned to a no-window room increases 800-fold, to 4 in 5, or 0.8.

The following provides an actual example (using the probabilities above) of a lawyer confusing the jury by ignoring additonal (Bayesian) information.

—JdP

[428]

The O.J. Simpson case... is more than tangentially germane to Bayes' Theorem, coincidences, and the relations between individual viewpoints and societal norms. [As an example of statisticide, consider] a refrain constantly repeated by attorney Alan Dershowitz during the trial. He declared that since fewer than 1 in 1,000 women who are abused by their mates go on to be killed by them, the spousal abuse in the Simpson's marriage was irrelevant to the case. Although the figures are correct, Mr. Dershowitz's claim is a stunning non-sequitur; it ignores the obvious fact: Nicole Simpson was killed. [Using Bayes' Theorem, it can be shown that among abused wives and girlfriends who are] later murdered, the batterer is the murderer more than 80 per cent of the time. Thus, *without any further evidence*, there was mathematical warrant for police suspicion of Mr. Simpson.

—John Allen Paulos *Once Upon a Number*,
Basic Books, 1998, pg. 74.,

[429]   Anyone who cannot cope with mathematics is not fully human. At best he is a tolerable subhuman who has learned to wear shoes, bathe and not make messes in the house.

—Robert Heinlein (1907–1988) spoken by Lazarus Long in
Time Enough for Love, Ace Books, 1994.

[430]   Mathematics takes us still further from what is human, into the region of absolute necessity, to which not only the actual world, but every possible world, must conform.

—Bertrand Russell (1872–1970) *The Study of Mathematics*, 1902.

Note: See related comments of Martin Gardner in a Platonist context, item **[675]**.

[431]   Whoever despises the high wisdom of mathematics, nourishes himself on delusion and will never still the sophistic sciences whose only product is an eternal uproar.

—Leonardo da Vinci (1452–1519)

[432]   Mathematics knows no physical bounds, [it] develops several critical habits of mind, [it] is particularly good for developing logic skills, [it] trains students to think abstractly, [and mathematics] teaches self-reliance in thinking.

—William J. Bennett, *The Educated Child*,
Simon & Schuster/Touchstone (2000), pp. 278–280.

[433]   Mathematics is the queen of the sciences and arithmetic is the queen of mathematics. She often condescends to render service to astronomy and other natural sciences, but under all circumstances the first place is her due.

—Carl Friedrich Gauss (1777–1855) quoted in:
S. von Waltershausen, *Gauss zum Gedächtnis* (1856).

MATHEMATICS IS AN ART

[434]

Pattern made by swinging pendulum during 2001 Seattle Earthquake
(c) 2001, Courtesy Norman MacLeod, Gaelic Wolf Consulting.

[435] Mathematics is the art of giving the same name to different things. . . . When the language is well chosen, we are astonished to learn that all the proofs made for a certain object apply immediately to many new objects; there is nothing to change, not even the words, since the names have become the same.

—Henri Poincaré (1854–1912) *Science and Method*, (essays), 1903.

[436] Poetry is the art of giving different names to the same thing.

—Anonymous

MATHEMATICS IS ECONOMICAL/NOT ECONOMICAL

[437] One merit of mathematics few will deny: it says more in fewer words than any other science. The formula, $e^{\pi} = -1$ expresses a world of thought, of truth, of poetry, and of religious spirit of "God eternally geometrizes."

—David Eugene Smith (1860–1944)
Professor of Mathematics, Columbia Teachers College
The Poetry of Mathematics and Other Essays

Note: See also the successful formula cartoon, item [217].

[438] Mathematics is the science which uses easy words for hard ideas.

—Edward Kasner (1878–1955) *Mathematics and the Imagination*,
New York: Simon and Schuster, 1940.

[439] In great mathematics there is a very high degree of unexpectedness, combined with inevitability and economy.

—Godfrey H. Hardy (1877–1947) *A Mathematician's Apology*,
London, Cambridge University Press, 1941.

[440] Mathematics is the cheapest science. Unlike physics or chemistry, it does not require any expensive equipment. All one needs for mathematics is a pencil and paper.

—Donald J. Albers and Gerald L. Alexanderson, ed.,
Mathematical People, Boston: Birkhäuser, 1985.

"...and thus, when mathematicians had collected unto themselves all the chalk in the Universe for their theorems, the chemists and the physicists said, 'Woe be unto us, for we have no chalk to meet our needs. Let us therefore create unto ourselves a first COMPUTER!' "

MATHEMATICS IS ETERNAL

[441] Perhaps the most surprising thing about mathematics is that it is so surprising. The rules which we make up at the beginning seem ordinary and inevitable, but it is impossible to foresee their consequences. These have only been found out by long study, extending over many centuries.

Much of our knowledge is due to a comparatively few great mathematicians such as Newton, Euler, Gauss, or Riemann; few careers can have been more satisfying than theirs. They have contributed something to human thought even more lasting than great literature, since it is independent of language.

—E.C. Titchmarsh (1899–1963)

[442] We could use up two Eternities in learning all that is to be learned about our own world and the thousands of nations that have arisen and flourished and vanished from it. Mathematics alone would occupy me eight million years.

—Mark Twain (1835–1910), born Samuel L. Clemens
Notebook #22, Spring 1883–Sept. 1884.

MATHEMATICS IS IMMUTABLE

[443]   In most sciences one generation tears down what another has built and what one has established another one undoes. In mathematics alone each generation adds a new story to the old structure.

—Hermann Hankel (1839–1873) Quoted in D. MacHale,
Comic Sections, Dublin 1993.

Note: See also item [580], "Is Progress Monotonically Increasing?" for the nature of scientific progress, and Martin Gardner's comment on the universality of mathematics, item [675].

[444]   Mathematicians discover equations that are timeless, as true tomorrow as they are today, and in so doing, often end up changing the future in irrevocable ways.

—Burt Nanus*, University of Southern California

MATHEMATICS IS LOGICAL: (DEDUCTIVE)

Note: For contrast, see item [337] in which Marston Morse minimizes the role of logic in mathematics.

[445]   It is true that mathematics, owing to the fact that its whole content is built up by means of purely logical deduction from a small number of universally comprehended principles, has not unfittingly been designated as the science of the self-evident (Selbstverständlichen).

Experience however shows, that for the majority of the cultured, even of scientists, mathematics remains the science of the incomprehensible (Unverständlichen).

—Alfred Pringsheim

[446]   The great science [mathematics] occupies itself at least just as much with the power of imagination as with the power of logical conclusion.

—Johann Friedrich Herbart (1776–1841), German philosopher
quoted in *Why Johnny Can't Add:
The Failure of the New Mathematics*,
St. Martin's Press, 1973, Ch. 4, by Morris Kline.

Note: See also "Logic and Reason," item **[319]**, and "Deduction vs. Induction," item **[160]**.

[447]   Mathematics is the science which draws necessary conclusions.

—Benjamin Peirce (1809–1880)
Memoir read before the National Academy of Sciences
in Washington, 1870.

Note: For more on "necessary" and "sufficient," see "Deduction vs. Induction," item **[160]**.

[448]

"You foolish witch! Forget Whitehead and his quantifiers! Mathematicians may be OK with 'any' things or 'some' things, but real toil and trouble require particular things!"

Mathematics as a science, commenced when someone, probably a Greek, proved propositions about "any" things or "some" things, without specifications of definite particular things.

—Alfred North Whitehead (1861–1947)

[449]   The art of doing mathematics consists in finding that special case which contains all the germs of generality.

—David Hilbert (1862–1943)
(For more on the roles of induction (generalization)
and deduction (special cases), see items [159] and [160].)

[450]   Certain characteristics of the subject are clear. To begin with, we do not, in this subject, deal with particular things or particular properties: we deal formally with what can be said about "any" thing or "any" property.

We are prepared to say that one and one are two, but not that Socrates and Plato are two, because in our capacity as logicians or pure mathematicians, we have never heard of Socrates or Plato. A world in which there were no such individuals would still be a world in which one and one are two.

It is not open to us, as pure mathematicians or logicians, to mention anything at all, because, if we do so we introduce something irrelevant and not formal.

—Bertrand Russell (1872–1970)
Introduction to Mathematical Philosophy,
ch. XVIII, Dover Pubns, October 1993.

[451]   [Mathematics] is that [subject] which knows nothing of observation, nothing of experiment, nothing of induction, nothing of causation.

—Thomas Henry Huxley (1825–1895)
The Scientific Aspects of Positivism, Fortnightly Review (1898);
Lay Sermons, Addresses and Reviews, (New York, 1872), pg. 169.

MATHEMATICS IS MADNESS/ADDICTIVE/IRRELEVANT

[452]   Let us grant that the pursuit of mathematics is a divine madness of the human spirit, a refuge from the goading urgency of contingent happenings.

—Alfred North Whitehead (1861–1947)
Science and the Modern World,
New York, The Macmillan Co., 1929.

[453]   I admit that mathematical science is a good thing. But excessive devotion to it is a bad thing.
> —Aldous Huxley (1894–1963) Interview with J. W. N. Sullivan,
> *Contemporary Mind*, London, 1934.

[454]   Pure mathematics consists entirely of assertions to the effect that, if such and such a proposition is true of anything, then such and such another proposition is true of that thing. It is essential not to discuss whether the first proposition is really true, and not to mention what the anything is, of which it is supposed to be true.

Both these points would belong to applied mathematics. We start, in pure mathematics, from certain rules of inference, by which we can infer that if one proposition is true, then so is some other proposition. These rules of inference constitute the major part of the principles of formal logic. We then take any hypothesis that seems amusing, and deduce its consequences. If our hypothesis is about anything, and not about some one or more particular things, then our deductions constitute mathematics.

Thus mathematics may be defined as the subject in which we never know what we are talking about, nor whether what we are saying is true.
> —Bertrand Russell (1872–1970) *Mysticism and Logic*, 1947, ch. 4.

[455]

> The math'matician is e'er among us
> We tolerate and feed 'em.
> We don't know what the heck they do
> We only know we need 'em.

> —John de Pillis

[456]   Mathematics is a game played according to certain simple rules with meaningless marks on paper.
> —David Hilbert (1862–1943) reflecting Hilbert's apprehension at
> the beginning of the 20th century when contradictions arose (e.g.,
> Russell's paradox) that challenged the foundations of mathematics.

MATHEMATICS IS SIMPLIFYING/UNIFYING

[457]   How thoroughly it is ingrained in mathematical science that every real advance goes hand in hand with the invention of sharper tools and simpler methods while at the same time, assist in understanding earlier theories and in casting aside some more complicated developments.

—David Hilbert (1862–1943)

[458] Mathematics, the science of the ideal, becomes the means of investigating, understanding and making known the world of the real. The complex is expressed in terms of the simple. From one point of view mathematics may be defined as the science of successive substitutions of simpler concepts for the more complex...

—William Frank White
A Scrap-Book of Elementary Mathematics, pg. 215.

[459]

"Of course he has changed, Virgil. It's the simplifying effect of all that mathematics he's studying."

"The essence of the mathematician's approach, and the key to its incredible power and success, lies in the extreme simplicity and highly abstract nature of the patterns singled out for study. Mathematicians take a highly simplistic view of the world, avoiding the complexities of nature and life. The mathematican views the world in terms of perfectly straight lines, perfect circles ..."

—Keith Devlin, Executive Director,
Center for the Study of Language and
Information, Stanford University
Life by the Numbers, Wiley, 1998, pg.18.

[460]   To most people, mathematics makes the world more complicated. It doesn't do that! Math makes the world more simple! Mathematicians are simplistic creatures. We look at the world in the simplest possible way. We look at it in such a simple way that the only way of capturing the simplicity is with symbols, lines, nodes, edges of graphs. We strip away the complexity!

–Keith Devlin, Executive Director,
Center for the Study of Language and
Information, Stanford University
Life by the Numbers television series,
episode 6 (PBS/WQED 1998)

[461]   Time was when all the parts of the subject were dissevered, when algebra, geometry, and arithmetic either lived apart or kept up cold relations of acquaintance confined to occasional calls upon one another; but that is now at an end; they are drawn together and are constantly becoming more and more intimately related and connected by a thousand fresh ties, and we may confidently look forward to a time when they shall form but one body with one soul.

—James Joseph Sylvester (1814–1897)
Presidential Address to the British Association, 1869.

[462]   The mathematical facts worthy of being studied are those which, by their analogy with other facts, are capable of leading us to the knowledge of a physical law. They... reveal the kinship between other facts, long known, but wrongly believed to be strangers to one another.

—Henri Poincaré (1854–1912)

[463]   Kepler's principal goal was to explain the relationship between the existence of five planets (and their motions) and the five regular solids. It is customary to sneer at Kepler for this.... It is instructive to compare this with the current attempts to "explain" the zoology of elementary particles in terms of irreducible representations of Lie groups.

—S. Sternberg, *Celestial Mechanics*,
W. A. Benjamin Inc., pg. 951, 1969.

[464]   The further a mathematical theory is developed, the more harmoniously and uniformly does its construction proceed, and unsuspected relations are disclosed between hitherto separated branches of the science.

—David Hilbert (1862–1943)

MATHEMATICS IS WIDE IN SCOPE

[465] I had a feeling once about Mathematics—that I saw it all. Depth beyond depth was revealed to me—the Byss and Abyss. I saw—as one might see the transit of Venus or even the Lord Mayor's Show—a quantity passing through infinity and changing its sign from plus to minus. I saw exactly why it happened and why the tergiversation was inevitable but it was after dinner and I let it go.

> —[Sir] Winston Spencer Churchill (1874–1965) quoted in
> *Return to Mathematical Circles*, by H. Eves, Boston: Prindle,
> Weber and Schmidt, 1988. (For another Churchill's
> (Lord Randolph) reaction to mathematics, see item [707].)

[466] Every new body of discovery is mathematical in form, because there is no other guidance we can have.

> —Charles Darwin (1809–1882)

[467] Pure and applied mathematics do not exist as separate entities—but the tensions between them do exist.

> —Anonymous

[468] Angling may be said to be so like mathematics that it can never be fully learned.

> —Izaak Walton (1593–1683) *The Complete Angler*
> (Everyman Paperback Classics). Author's Preface.

[469] It is true that Fourier had the opinion that the principal aim of mathematics was public utility and explanation of natural phenomena; but a philosopher like him should have known that the sole end of science is the honor of the human mind, and that under this title, a question about numbers is worth as much as a question about the system of the world.

> —Karl Jacobi (1804–1851)

[470] Mathematics has a triple end. It should furnish an instrument for the study of nature. Furthermore, it has a philosophic end, and, I venture to say, an end esthetic.

> —Henri Poincaré (1854–1912)

MATHEMATICS IS YOUTHFUL

[471]   Mathematics is like checkers in being suitable for the young, not too difficult, amusing, and without peril to the state.

—Plato (ca 429–347 B.C.)

[472]   Mathematics is one of the oldest of the sciences; it is also one of the most active, for its strength is the vigour of perpetual youth.

—A.R. Forsyth (1858–1942)

[473]   It is the perennial youthfulness of mathematics itself which marks it off with a disconcerting immortality from the other sciences.

—Eric Temple Bell (1883–1960)

METHOD

Time Management

[474]

How do you find enough time to do all the things you want to do?

Well, there are 24 hours in every day.... And, if that's not enough, you've always got the nights.

—Ron Graham*, University of California, San Diego

Test for a Mathematician

[475]   Some years ago, a little problem was proposed to ferret out potential mathematicians by their method of reasoning.

The person was asked to imagine being in a kitchen containing a gas stove with one burner lit, and also containing a kettle of water on the floor. The person was then asked how to heat the water in the kettle. Naturally the usual response was to place the kettle on the lit burner.

Then a second problem was posed; this problem was identical to the first except that the kettle was now on the kitchen table.

If the person responded that the kettle should be taken from the table and placed on the lit burner, the person had no chance of becoming a mathematician. For a real mathematician would transfer the kettle from the table to the floor, thereby reducing the second problem to the first which had already been solved.

—Anonymous

[476]   What is the difference between a method and a device? A method is a device which you use twice.

—George Pólya
How to Solve It, pg. 181.

[477]

It is virtually impossible to get anything exactly right.

—Carl de Boor*, University of Wisconsin

Metric Maxims

[478]

28.3 grams of prevention are worth .453 kilograms of cure.
Give him 2.5 centimeters and he'll take 1609 meters.
More bounce to the 28.3 grams.
He demanded his .453 kilograms of flesh.
155 centimeters and eyes of blue.
A miss is as good as 1.6 kilometers.
First down and 9.1 meters to go.
I love you 35.2 liters and 8.8 liters.

—Anonymous

Möbius Strip

[479]

Snip Snip

The topologist's mind came unguided
When his theories, some colleagues derided.
    Out of Möbius strips
    Paper dolls he now snips,
Non-Euclidean, closed and one-sided.

—H. Schenk, Jr., *Fantasy and Science Fiction*,
Mercury Press, Sept. 1959

[480]

A mathematician confided
That a Möbius strip is one-sided.
    And you'll get quite a laugh
    if you cut one in half,
For it stays in one piece when divided.

—Cyril Kornbluth (1923–1958), U.S. Writer

N

NATURAL BASE

[481]

Mathematical Constance

I think that I shall never see
A constant lovelier than e.

Whose digits are too great to state;
They're 2.71828...

And e has such amazing features.
It's loved by all (but mostly teachers).

And Calculus, we'd not do well in
Without such terms as exp and ln.

For *e* has such nice properties,
Most integrals are done with...ease.

Theorems are proved by fools like me,
But only Euler could make an *e*.

—Arthur Benjamin, Harvey Mudd College
The American Mathematical Monthly, vol. 108, no. 5,
May 2001, pg. 423 (from a tribute to Constance Reid,
with apologies to Joyce Kilmer)

NATURAL NUMBERS

[482]   Once, in a taxi from London, Hardy noticed its number, 1729. He must have thought about it a little because he entered the room where Srinivasa Ramanujan (1887–1920) lay in bed and, with scarcely a hello, blurted out his disappointment with it. It was, he declared, "rather a dull number," adding that he hoped that wasn't a bad omen. "No, Hardy," said Ramanujan, "it is a very interesting number. It is the smallest number expressible as the sum of two cubes in two different ways."

—As told by G.H. Hardy, from *The Man Who Knew Infinity*,
by Robert Kanigel, Washington Square Press, Rep. edition (1991).

[483]

The number 1729 has since become known as the Hardy-Ramanujan number, although this property of 1729 was known more than 300 years before Ramanujan. Numbers of its type (the smallest numbers expressible as the sum of 2 cubes in n ways) are sometimes called Taxicab Numbers.

—*JdP*

[484]   [Leopold Kronecker (1823-1891)] would have rid mathematics of the articial numbers and of its "symbolic" methods, and the devising of new functons seemed to him a foolish waste of energy. "God created numbers and geometry," I once heard him say, "but man the functions."

—Henry Fine, "Kronecker and His Arithmetical Theory
of the Algebraic Equations," *Bull. Amer. Math. Soc.*,
1892, No. 1, pg 183.

[485]   Number rules the universe.

—Pythagoras (ca. 582–507 B.C.)

[486]   The numbers are a catalyst that can help turn raving madmen into polite humans.

—Philip J. Davis, Emeritus Professor,
Applied Mathematics, Brown University.

NEWTON

[487]

> When Newton saw an apple fall, he found
>   In that slight startle from his contemplation—
> T'is said (for I'll not answer above ground
>   For any sage's creed or calculation)—
> A mode of proving that the earth turn'd round
>   In a most natural whirl, called 'gravitation';
> And this is the sole mortal who could grapple,
>   Since Adam, with a fall or with an apple.

—Lord George Gordon Byron (1788–1824)
Canto the Tenth of Don Juan

Note: See also comments of Charles Darwin on the challenge of understanding our place in the Universe, item [212].

[488]   Taking mathematics from the beginning of the world to the time of Newton, what he has done is much the better half.

—Gottfried Wilhelm Leibniz (1646–1716) Quoted in
A History of Mathematics
by C.B. Boyer, (New York 1968)

[489]   I recognize the lion by his paw.

—Jakob Bernoulli (1654–1705) After reading an
anonymous solution to a problem that he realized was
Newton's solution. In G. Simmons, *Calculus Gems*,
New York: McGraw-Hill, 1992, pg. 136.

[490]   The law of gravitation is indisputably and incomparably the greatest discovery ever made, whether we look at the advance which it involved, the

Sir Isaac Newton

extent of truth disclosed, or the fundamental and satisfactory nature of this truth.

—W. Whewell (1794–1866)

Note: See Heilbron's, item **[594]**, for a history of the art of deduction, item **[160]** for a discussion of "Deduction vs. Induction" and item **[578]** for "Origins of scientific method."

[491]   There may have been minds as happily constituted as his for the cultivation of pure mathematical science; there may have been minds as happily constituted for cultivation of science purely experimental; but in no other mind have the demonstrative faculty and the inductive faculty co-existed in such supreme excellence and harmony.

—Lord Macaulay, also Thomas Babington, (1800–1859)

[492]   On the day of Cromwell's death, when Newton was sixteen, a great storm raged all over England. He used to say, in his old age, that on that day he made his first purely scientific experiment. To ascertain the force of the wind, he first jumped with the wind and then against it, and, by comparing these distances with the extent of his own jump on a calm day, he was able to compute the force of the storm. When the wind blew thereafter, he used to say it was so many feet strong.

—James Parton

[493]   This almost superhuman genius, whose powers and attainments at once make us proud of our common heritage, and humble us with our disparity.

—Thomas Brown

[494] The efforts of the great philosopher (Newton) were always super-human; the questions which he did not solve were incapable of solution in his time.

—François Arago (1786–1853)

"Don't blame me, Mom. Newton invented gravity!"

[495]

Nature and Nature's laws lay hid in the night;
God said, "Let Newton be" and all was light.

—Alexander Pope (1688–1744)

Note: This couplet of Alexander Pope is the engraved epitaph on Isaac Newton's sarcophagus in Westminster Abbey, London.

[496]

It did not last: the Devil howling,
  "Ho! Let Einstein be!"
Restore the status quo.

—J.C. Squire (1884–1958) English parodist
Reply to previous entry, Pope's epitaph on Isaac Newton

[497]

While studying pressures and suctions,
Sir Isaac performed some deductions,

"Fill a mug to the brim, it
Will then reach a limit,
So clearly determined by fluxions.

—Paul Ritger

[498]

I'm not smart, but I like to observe. Millions saw the apple fall, but Newton was the one to ask why.

—Bernard Baruch (1870–1965)

[499]   Newton was the greatest genius that ever existed, and the most fortunate, for we cannot find more than once a system of the world to establish.

—Joseph-Louis Lagrange (1736–1813)

[500]   I don't know what I may seem to the world, but, as to myself, I seem to have been only as a boy playing on the seashore, and diverting myself in now and then finding a smoother pebble or a prettier shell than ordinary, whilst the great ocean of truth lay all undiscovered before me.

—Sir Isaac Newton (1642–1726)

[501]   If I have seen farther than Descartes, it is by standing on the shoulders of giants.

—Sir Isaac Newton (1642–1726)

[502]   Newton could not admit that there was any difference between him and other men, except in the possession of such habits as perseverance and vigilance. When he was asked how he made his discoveries, he answered, "by always thinking about them;" and at another time he declared that if he had done anything, it was due to nothing but industry and patient thought: "I keep the subject of my inquiry constantly before me, and wait till the first dawning opens gradually, by little and little, into a full and clear light."

—W. Whewell (1794–1866)

[503]

> O'er Nature's laws God cast the veil of night,
> Out-blaz'd a Newton's soul—and all was light.

—Aaron Hill (1685–1750)

NORMAL SPACE

[504]

"Look at that degenerate matrix!"

Normal Space: This nomenclature is an excellent example of the time-honored custom of referring to a problem we cannot handle as *abnormal, irregular, improper, degenerate, inadmissable* and otherwise undesirable.

—John Kelley (1916–1999)

OBVIOUS

[505] "Obvious" is the most dangerous word in mathematics.

—Eric Temple Bell (1883–1960)

[506] Obviousness is always the enemy to correctness. Hence we invent some new and difficult symbolism, in which nothing seems obvious. Then we set up certain rules for operating on the symbols, and the whole thing becomes mechanical.

—Bertrand Russell (1872–1970)

[507] A thing is obvious mathematically—after you see it.

—Robert Daniel Carmichael (1879–1967) (For the role of Carmichael numbers relative to Fermat's little theorem and the probablistic test for prime numbers, see item **[606]**.)

[508] Not seldom did he [Sir William Thomson (Lord Kelvin)] in his writing set down some mathematical statement with the prefacing remark "it is obvious that" to the perplexity of mathematical readers, to whom the statement was anything but obvious from such mathematics as preceded it on the page. To him it was obvious for physical reasons that might not suggest themselves at all to the mathematician, however competent.

—S. P. Thompson (1851–1916) *The Life of William Thomson Baron Kelvin of Largs*

OFFENSIVE AND UNPOPULAR MATHEMATICS

[509]   And who can doubt that it will lead to the worst disorders when minds created free by God are compelled to submit slavishly to an outside will? When we are told to deny our senses and subject them to the whim of others? When people devoid of whatsoever competence are made judges over experts and are granted authority to treat them as they please? These are the novelties which are apt to bring about the ruin of commonwealths and the subversion of the state.

—Galileo Galilei (1564–1642)
In James R. Newman (ed.) *The World of Mathematics*,
New York: Simon and Schuster, 1956, pg. 733.

[510]   I remember one occasion when I tried to add a little seasoning to a review, but I wasn't allowed to. The paper was by Dorothy Maharam, and it was a perfectly sound contribution to abstract measure theory. The domains of the underlying measures were not sets but elements of more general Boolean algebras, and their range consisted not of positive numbers but of certain abstract equivalence classes. My proposed first sentence was: "The author discusses valueless measures in pointless spaces."

—Paul Halmos, In: *I Want to Be a Mathematician*,
Washington: MAA Spectrum, 1985, pg. 120.

[511]   There is no national science just as there is no national multiplication table; what is national is no longer science.

—Anton Chekov, (1860–1904)
In V.P. Ponomarev, *Mysli o nauke Kishinev*, 1973.

P

PARADOX

Note: See also item **[64]** for a weak form of the **Banach-Tarski** paradox with an accessible proof that only requires knowledge of (countable) integer cardinality and (uncountable) real number cardinality.

[512] One of the endlessly alluring aspects of mathematics is that its thorniest paradoxes have a way of blooming into beautiful theories.

> —Philip J. Davis, Number, *Scientific American*, 211, (Sept. 1964), 51–59.

[513] Perhaps the greatest paradox of all is that there are paradoxes in mathematics.

> —Edward Kasner and James R. Newman

[514]

> FREDERIC: *(more amused than angry)*
> How quaint the ways of Paradox!
> At common sense she gaily mocks!
> Though counting in the usual way,
> Years twenty-one I've been alive,
> Yet, reck'ning by my natal day,
> Yet, reck'ning by my natal day,
> I am a little boy of five!
> —William S. Gilbert, librettist, music by Sir Arthur Sullivan
> *Pirates of Penzance*

Note: See also items [242]–[243], and [747] for more Gilbert and Sullivan parodies.

[515]   Not truth, not certainty. These I foreswore in my novitiate, as young men called to holy orders must abjure the world.

"If..., then...,"

this only I assert; and my successes are but pretty chains linking twin doubts, for it is vain to ask if what I postulate be justified, or what I prove possess the stamp of fact.

Yet bridges stand, and men no longer crawl in two dimensions. And such triumphs stem, in no small measure, from the power this game, played with the thrice-attenuated shades of things, has over their originals.

How frail the wand, but how profound the spell.

—Clarence R. Wiley, Jr.

[516]   As lightning clears the air of impalpable vapours, so an incisive paradox frees the human intelligence from the lethargic influence of latent and unsuspected assumptions. Paradox is the slayer of prejudice.

—James Joseph Sylvester (1814–1897)

The Amazing Banach-Tarski Paradox

Note: See also item [50] for a discussion of the Axiom of Choice and items [55]–[64] to find a proof of a weak form of the Banach-Tarski Paradox.

[517]   The Banach-Tarski theorem must surely be the weirdest result that the mathematicians of the world have come up with. In a somewhat naïve form it states that we can take a solid ball, cut it into a finite number of pieces and then rearrange the pieces to create two solid balls—each the same size as the original one! This theorem is so absurd, that it is commonly referred to as the Banach-Tarski paradox.

This sounds absolutely preposterous and of course this is nothing that is physically possible. However, using the somewhat suspicious Axiom of Choice, this can be done mathematically!

—Frank Wikström, seminar notes, February, 1995.
Umeå universitet, Sweden.

[518]   Response to Wikström: No wonder the Axiom of Choice is controversial! Who wants an admittedly "suspicious" (if not paranoid) axiom lurking about our classrooms?

—Gerald L. Alexanderson*, Santa Clara University

[519]   But Tarski was more than just a logician. . . . His 1924 theorem with fellow Pole Stefan Banach—that one can divide a solid sphere into a finite number of "pieces" (five pieces actually suffice, as was shown later) and then put the pieces together again to form two solid spheres, each of the same size as the original one—became known as the Banach-Tarski Paradox and illuminates limitations of any mathematical theory of volume applying to all "pieces" of space. An irate citizen once demanded of the Illinois legislature that they outlaw the teaching of this result in Illinois schools!

—J. P. Addison, from an obituary of Alfred Tarski in
California Monthly, the UC Berkeley alumni magazine.

PHYSICISTS

[520]   Physicists know what is important but they don't know what is true. Mathematicians know what is true but they don't know what is important.

"OH, C'MON, MITTENS, HOW SMART CAN HUMANS BE
IF THEY THINK THAT STRINGS ARE DIFFICULT?"

[521]   Physicists often resort to the principle of explaining a general phenomenon by the statistical regularities that exist among irregularities.

—C.S. Peirce (1839–1914)

PI

[522]   Value of pi to the first 500 decimal places:

$= 3.$ 1415926535 8979323846 2643383279 5028841971 6939937510
5820974944 5923078164 0628620899 8628034825 3421170679
8214808651 3282306647 0938446095 5058223172 5359408128
4811174502 8410270193 8521105559 6446229489 5493038196
4428810975 6659334461 2847564823 3786783165 2712019091
4564856692 3460348610 4543266482 1339360726 0249141273
7245870066 0631558817 4881520920 9628292540 9171536436
7892590360 0113305305 4882046652 1384146951 9415116094
3305727036 5759591953 0921861173 8193261179 3105118548
0744623799 6274956735 1885752724 8912279381 8301194912

[523]   It can be of no practical use to know that Pi is irrational, but if we can know, it surely would be intolerable not to know.

—E. C. Titchmarsh (1899–1963)

Mnemonics for the Digits of Pi

[524]

How I want a drink, alcoholic of course, after the heavy lectures
 3   1   4   1   5   9   2   6   5   3   5   8

involving quantum mechanics.
 9   7   9

—Jan Lindheim (Cal. Inst. of Technology)
as learned in English, in a Norwegian prep school (gymnas)

[525]

May I have a large container of coffee — sugar and cream?
 3   1   4   1   5   9   2   6   5   3   5

—Anonymous

[526]

Now, I even I, would celebrate
 3   1   4   1   5   9

In rhymes inept, the great
 2   6   5   3   5

immortal Syracusan, rivaled nevermore
 8   9   7   9

Who in his wondrous lore, passed on before
 3   2   3   8   4   6   2   6

Left men his guidance how to circles mensurate.
 4   3   3   8   3   2   7   9

—A. C. Orr

[527]

Riddle: Why can't there be a mnemonic, such as those above which yields the first 33 digits of Pi?

Hint: What is the 33rd digit?

[528]

Re Pi, (No. 1)

A man in an ivory tower
Dreamed of pi to the 200th power.
        (This is apt to occur
        when you mix a liqueur
With clams at a very late hour.)

[529]

Re Pi, (No. 2)

A Biblical version of pi
Is recorded by some unknown guy
    In "Kings"* where he mentions
    A column's dimensions
Not exact but a pretty good try.

        —John McClellan, *The Mathematics Teacher*,
        vol. 59, no. 7, 1966. *Kings 1, 7:23

PLAY

[530] The problem with juggling is that the balls go where you throw them. Just as the problem with programming is that the computer does exactly what you tell it.

        —Ron Graham*, University of California, San Diego.
        On parallels between juggling and computing.

[531]   Can it be that all great scientists of the past were really playing a game, a game in which the rules are written not by man but by God?

When we play we do not ask why we are playing—we just play. Play serves no moral code except that strange code which for some unknown reason, imposes itself on the play ...

"When you finish browsing in there, Buford, come outside and browse with the rest of us."

You will search in vain through scientific literature for hints of motivation. And as for the strange moral code observed by scientists, what could be stranger than an abstract regard for truth in a world which is full of concealment, deception, and taboos?

In submitting to your consideration the idea that the human mind is at its best when playing, I am myself playing, and that makes me feel that what I am saying may have in it an element of truth.

—J. L. Synge (1897–1995)

PRIME NUMBERS

[532]

Definition: p > 1 is a prime number if p is not divisible by any positive integer except 1 and itself.

Tests for primality: To see whether an integer is prime or not, probability theory can be applied. Numbers that pass the probability tests are declared to be prime—but only to a very high-probability. (Is this a "proof?") For large integers used in encryption, the probablistic test is practical and convenient. For more on this point, see item **[607]**.

—JdP

[533]

I can't cut this steak, he confided
To the waiter who simply recited,
    Your prime cut, of course
     Is as tough as a horse
Since you can't take a prime and divide it.

—John de Pillis

Mersenne Primes

[534] **Cartoon Note:** As of this writing, $2^{(13,466,917)} - 1$, which requires 4,053,946 digits to write, is the largest known prime. On November 14, 2001, **Michael Cameron**, 20, from Canada, discovered this prime (the 39th

"Lil' Fermat may say he's only <u>probably</u> prime, but his Smith 'n Wesson sez he's a prime fer <u>sure!</u>"

Mersenne prime) using software from the Great Internet Mersenne Prime Search (GIMPS). GIMPS is an organization that provides free software and source code for some 130,000 personal computer users wishing to find ever bigger prime numbers.

[535]

*Definition: A **Mersenne prime** is a prime of the form $2^P - 1$ where p is also prime. The first Mersenne primes are $3 = 2^2 - 1$, $7 = 2^3 - 1$, $31 = 2^5 - 1$, $127 = 2^7 - 1$, etc. As of November 14, 2001, there were only 39 known Mersenne primes.*

—JdP

[536]   Proof that All Odd Numbers Are Prime

- **The Mathematician's Proof:** Let's see, 1 is prime, 3 is prime, 5 is prime, and 7 is prime. The result follows by induction.

- **The Physicist's Proof:** OK, 3 is prime, 5 is prime, 7 is prime, 9 is not prime (Oops, experimental error!), 11 is prime, 13 is prime, and so on.

- **The Engineer's Proof:** 3 is prime, 5 is prime, 7 is prime, 9 is prime, 11 is prime, 13 is prime, 15 is prime, 17 is prime, etc.

Moral: Just because there are many roads to the mountain top, doesn't mean you are on the right mountain.

—*JdP*

PRINCIPLES IN MATHEMATICS

[537]   Principle of Conservation of Ignorance   A false notion once arrived at is not easily dislodged.

—Georg Cantor (1845–1918)

[538]   Principle of the Existence of Heaven and Hell   God exists because mathematics is consistent—the devil exists because we cannot prove the consistency.

—Hermann Weyl (1885–1955)

[539]   The Reflection Principle   In most teaching the ideas just bounce back, without penetration.

—Peter Ross, Santa Clara University

[540]   Problem Solving Principle   If you have a problem you cannot solve, change it into one that you can solve. In other words, if you aren't near the one you love, love the one you're near.

[541]   Principle of Maximum Simplicity   Everything should be as simple as possible but no simpler.

—Albert Einstein (1879–1955)
Reader's Digest, Oct. 1977.

Note: See also *Occam's Razor*, item **[189]**: Simplicity as a guideline.

[542]

The Pigeon Hole Principle

"In ten pigeon holes," said Dirichlet,
"Eleven fat pigeons waited."
The master then went on to say,
"At last two of them are mated."

—James P. Burling
The American Mathematical Monthly, vol. 72, no. 5, 1966, pg. 527.

[543]   **Principle of Least Astonishment**   A theorem is true if everyone would be astonished if it were false.

[544]   **Principle of Impotence**   One cannot devise a successful gambling system that can be used against a fair coin.

[545] Principle of Non-reciprocity of Expectation Negative expectations yield negative results; positive expectations yield negative results.

[546] Principle of Invariance of Grades Professors constantly try to make courses more appropriate, more relevant, more interesting, more intelligible to the students, yet the distribution of grades remains amazingly unaffected by these efforts.

[547] Principle of Static Queuing Theory The other line is always shorter.

[548] Principle of Dynamic Queuing Theory The other line always moves faster.

[549] Penultimate Principle Everything is a special case of something more general.

[550] Ultimate Principle This principle is so perfectly general that no specific application of it is possible.

PROBABILITY

[551] Stephen Jay Gould on the Awesome Continuity of Life

And within the smaller, but still tolerably ample, compass of our planetary real estate, I would nominate as most worthy of pure awe—a metaphorical miracle, if you will—an aspect of life that most people have never considered but that strikes me as equal in majesty to our most spiritual projections of infinity and eternity—the continuity of *etz chayim*, the tree of earthly life, for at least 3.5 billion years, without a single microsecond of disruption.

Consider the improbability of such continuity in the conventional terms of ordinary probability: take any phenomenon that begins with a positive value at its inception 3.5 billion years ago, and let the process regulating its existence proceed through time. A line marked zero runs along below the current value. The probability of the phenomenon's descent to zero may be almost incalculably low, but throw the dice of the relevant process billions of times and the phenomenon just *has* to hit the zero line eventually.

"C'mon, now, Earl. You'd think a little asteroid shower here in the Yucatan is the worst thing in the world. That's no excuse not to clean the ptericoccolis bones out of the garage!"

Cartoon Note: The role of the Yucatan peninsula in the extinction of dinosaurs is discussed in item **[168]**.

For most processes, the prospect of such an improbable crossing bodes no permanent ill, because an unlikely crash (a year, for example, when a healthy Mark McGwire hits no home runs at all) will quickly be reversed, and ordinary residence well above the zero line re-established. But life represents a different kind of ultimately fragile system, utterly dependent upon unbroken continuity. For life, the zero line designates a permanent end, not a temporary embarrassment. If life had ever touched that line for one fleeting moment at any time during 3.5 billion years of sustained history, neither we nor a million species of beetles would grace this planet today. A single handshake with voracious zero dooms all that might have been, forever after.

Well, I may be a rationalist at heart, but if anything in the natural world merits designation as "awesome," I nominate the continuity of the tree of life for 3.5 billion years.

—Stephen Jay Gould
American Museum of Natural History site
http://staging.amnh.org/naturalhistory/features/1200_feature.html,
Dec 2000/Jan 2001.

Note: For more on the extinction of dinosaurs, see item [**168**].

[**552**]   A reasonable probability is the only certainty.
—Edgar Watson Howe (1853–1937), American journalist
Country Town Sayings, 1911.

[**553**]   It is an old maxim of mine that when you have excluded the impossible, whatever remains, however improbable, must be the truth.
—Sir Arthur Conan Doyle (1859–1930)
spoken by Sherlock Holmes in *The Adventure of Beryl Coronet*

[**554**]   To sum up, we can ascertain that, approximately, the frequency of an event is to the number of all observations as the probability of the event is to the probability of certainty, i.e. to one. I find this correspondence between facts and logic, between possibility and realization, wonderful, indeed!
—Blaise Pascal (1623–1662)

[**555**]

The laws of probability are so cleverly formulated, that it is impossible to devise an experiment to test their validity.

[556]   The excitement that a gambler feels when making a bet is equal to the amount he might win times the probability of winning it.

—Blaise Pascal (1623–1662)

[557]   It is remarkable that this science (probability), which originated in the consideration of games of chance, should have become the most important object of human knowledge.

—Pierre-Simon de Laplace (1749–1827)

[558]

Lest men suspect your tale untrue,
Keep probability in view.

—John Gay (1685–1732)

Fables. Part I. The Painter who Pleased Nobody and Everybody.

[559]   He who has heard the same thing told by 12,000 eye witnesses has only 12,000 probabilities, which equals one strong probability, which is far from certainty.

—Voltaire (1694-1778)

[560]   Is probability probable?

—Blaise Pascal (1623–1662)

[561]   Heisenberg was here ... perhaps.

Note: See also item **[130]** for more on W. K. Heisenberg.

[562]   It is a truth very certain that, when it is not in our power to determine what is true, we ought to follow what is most probable.

—René Descartes (1596–1650)

[563]   If a coin is tossed a hundred times and comes up heads each time, those ignorant of probability will say the "law of averages" will practically guarantee that it comes up tails the next time. Any one who knows the least bit of probability knows that on the next toss, the probability of coming up heads is 1/2 and coming up tails is also 1/2. (Unless the coin is double-headed!)

"Wait till you see the other side!"

[564]   Pascal's Wager: Pascal used probability to validate belief in God. He reasoned that no matter how low the probability might be for the existence of God and heaven, one should bet on God's existence since the payoff is infinite everlasting life, as opposed to everlasting damnation. The "expected value" of the wager favoring God's existence is therefore infinitely great.

Note: See also "Faith, Religion, and Spirituality," item **[564]**.

PROBABILITY VS. DETERMINISM

[565]   Billiard balls ... are routinely invoked as the model of predictable physics. No one calls billiards a game of chance. All it requires to get the ball into the pockets is a well-measured poke of the perfectly positioned cue.

"Have a heart, Melvyn. If you let him play he'll only roll snake eyes!"

Dice, on the other hand, [are perceived to be] the epitome of chance.
Yet, billiard balls and dice are both governed by the same physical laws.

—K. C. Cole, science writer
The Universe and the Teacup, Harcourt Brace & Co., 1998, pg. 132.

Note: See also item **[768]** for George Washington's comments on gambling.

[566]   So in the final analysis, chance lies in the clumsiness, the inexperience, or the naiveté of the thrower—or in the eye of the observer. In fact, one might perfectly imagine a civilization in which the rolling of the dice would be a sport and billiards a game of chance.

—Ivar Ekeland
The Broken Dice, Chicago, The University of Chicago Press, 1993.

[567]

*Mathematical functions are deterministic or predictable
in the sense that the same input always produces the same
output. This is something of a philosophical stumbling
block when (necessarily deterministic) functions are used
to generate "**random**" numbers. Hence, the more honest
appelation, "pseudo-random," is often used for these
numbers. Here follows John von Neumann's comment.*

—*JdP*

Anyone who considers deterministic (or arithmetical) methods of producing random digits is, of course, in a state of sin.

—John von Neumann (1951), quoted in Knuth, Donald
The Art of Computer Programming, Addison Wesley, 1968, Vol. 2;
also in Goldstine, Herman H., *The Computer from Pascal to von
Neumann*, Princeton Univ. Press, 1972, pg. 297.

PROBLEM SOLVING

[568]   Mark Twain Encounters a Word Problem from Arithmeticus

(the problem)   If it would take a cannon ball $3\frac{1}{8}$ seconds to travel four
miles, and $3\frac{3}{8}$ seconds to travel the next four, and $3\frac{5}{8}$ seconds to travel the

next four, and if its rate of progress continued to diminish in the same ratio, how long would it take the cannon ball to go fifteen hundred miles?

—from *Arithmeticus*

(the answer)   I don't know.

—Mark Twain (1835–1910), born Samuel L. Clemens
The Celebrated Jumping Frog of Calaveras County, and Other Sketches (New York: C. H. Webb, 1867; BoondocksNet Edition, 2001). http://www.boondocksnet.com/twaintexts/frog/ (July 7, 2001).

[569]   Each problem that I solved became a rule which served afterwards to solve other problems.

—René Descartes (1596–1650)
Discours de la Méthode. 1637.

[570]   I keep the subject constantly before me and wait till the first dawnings open little by little into full light.

—Sir Isaac Newton (1642–1726)

[571]

Professionals in all areas are expected by their clients to exhibit one quality they are supposed to have in common—sound judgment. The following story provides an illustration.

—JdP

In order to solve a persistent and intractable problem with one of its new multi-million dollar machines, a company contacted one of its recently retired engineers.

After studying the machine for a few hours, the engineer placed a small "x" in chalk on a particular component of the machine and proudly stated, "This is where your problem is. Replace this part." The part was replaced and the machine worked perfectly again. The company then received a bill for $50,000 from the engineer for his service.

Feeling somewhat put upon, the company demanded an itemized accounting of the engineer's charges. The engineer's response read: One chalk mark $1.00. Knowing where to put it, $49,999.00.

—Anonymous

[572] Don't just read it; fight it! Ask your own questions, look for your own examples, discover your own proofs. Is the hypothesis necessary? Is the converse true? What happens in the classical special case? What about the degenerate cases? Where does the proof use the hypothesis?

—Paul Halmos
I Want to Be a Mathematician, Washington: MAA Spectrum, 1985.

[573] When you're studying something new and feel confused, it's not bad—it's good! When you're really confused, you don't feel confused. True confusion is undetectable.

Larry, confused over the notion of "mouse," forges ahead with utmost confidence.

When you feel confused, it means your brain is in the process of straightening things out and becoming less confused.

—John Baez*, University of California, Riverside

[574] It is quite a three-pipe problem, and I beg that you won't speak to me for fifty minutes.

—Sir Arthur Conan Doyle (1859–1930)
spoken by Sherlock Holmes in *The Red-Headed League*

PROGRESS

[575] The more progress physical sciences make, the more they tend to enter the domain of mathematics, which is a kind of centre to which they all converge. We may even judge of the degree of perfection to which a science has arrived by the facility with which it may be submitted to calculation.

—Adolphe Quetelet (1796–1874) Belgian mathematician,
astronomer, statistician, and sociologist known for applying
statistics and probability to social phenomena.

[576] Nothing has more retarded the advancement of learning than the disposition of vulgar minds to ridicule and vilify what they cannot comprehend.

—Samuel Johnson (1709–1784)
The Rambler (essay #117 in three volumes)
Thomas Tegg, London, 1826.

[577] The reasonable man adapts himself to the world, the unreasonable man persists in trying to adapt the world to himself. Therefore, all progress depends on the unreasonable man.

—George Bernard Shaw (1856–1950)

Note: See also item **[471]** for Eugene Wigner's "complaint" about the unreasonable effectiveness of mathematics.

*"Our old methods won't work, Mugsy.
It keeps asking me for a password!"*

[578]   Origins of scientific method Every civilization of which we
have records has possessed a technology, an art, a religion, a political system, laws and so on. . . . But only the civilizations that have descended from
Hellenic Greece have possessed more than the most rudimentary science.
The bulk of scientific knowledge is a product of Europe in the last four centuries. No other place and time has supported the very special communities
from which scientific productivity comes.

—Thomas Kuhn (1922–1996),
The Structure of Scientific Revolutions, University of Chicago Press,
pp. 167–168, 3rd edition 1996.

Is Progress Monotonically Increasing?

[579]   Any jackass can kick down a barn, but it takes a carpenter to build
one.

—Sam Rayburn (1882–1961),
(Member, U.S. House of Representatives, 1913–1961,
Speaker, 1940–47; 1949–53; 1955–61.)

[580]

*Scientists and Non-scientists on "Progress": In the
remarks below, Thomas Kuhn claims that the practicing
scientist suffers from a myopic vision that distorts history.
Kuhn suggests that the scientist cheerfully assumes that
scientific progress always increases.*

*But instead of relying on what person A (Kuhn) tells
person B (the reader) what person C (the scientist) is
thinking, we can go directly to the sources:*

*First, there are Kuhn's remarks (item [581]) followed by
Alex Levine's endorsement (item [582]) of Kuhn's
assertion that scientists, such as physicist and Nobel
laureate, Steven Weinberg, falsely assume that science
always leads to the truth. Then there is the reply (item
[583]) of Steven Weinberg himself. See also, the relevant
comments of Max Planck, item [584].*

—JdP

Note: See also "Definitions for the Non-Scientist," item **[77]**, for a discussion of Kuhn's role among non-scientists as a chief spokesman for scientists and how deeply committed he is to the use of clear definitions.

[581]   Kuhn's view of scientists' view of progress   Scientific education makes use of no equivalent for the art museum or the library of classics, and the result is a sometimes drastic distortion in the scientist's perception of his discipline's past. More than the practioners of other creative fields, he comes to see it as leading in a straight line to the discipline's present vintage. In short, he comes to see it as progress.

—Thomas Kuhn (1922–1996)
The Structure of Scientific Revolutions, University of Chicago Press,
pg. 167, 3rd edition, 1996.

[582]   Alex Levine's support of Kuhn   [One] point of contention [among scientists] is Kuhn's view of scientific progress, [which] guarantees scientific change, but fails to guarantee that science inevitably draws closer to the truth.

According to [a recent article of Steven] Weinberg, "All this is wormwood to scientists like myself, who think the task of science is to bring us closer and closer to objective truth."

It is one thing to construe the task of science in this way. It is quite another to assume, as Weinberg appears to, that science must inevitably succeed in this pursuit. The view of the history of human knowledge as some grand triumphal procession toward the distant temple of truth is, it seems to me, profoundly unscientific.

—Alex Levine, Department of Philosophy, Lehigh University
(In support of Kuhn's view on how scientists view progress)
New York Review of Books, Letters to the Editors, T.S. Kuhn's
Non-Revolution: An Exchange, February 18, 1999.

[583]   Steven Weinberg's Reply   I never argued that progress toward objective truth is inevitable.... I think that such progress has, in fact, happened again and again in the history of modern science. Our future progress toward a fundamental physical theory may be stopped if it turns out that humans are not smart enough to conceive of such a theory (which I doubt), or if we find an infinite regress of more and more fundamental theories (also unlikely), or if society stops providing the resources for continued experiments (all too possible). But although I can't claim to know that we will be able to continue our progress toward a simple, objectively true theory underlying all natural phenomena, I do think that we must act under this assumption, for if we do not then our progress will surely not continue.

—Steven Weinberg, (Nobel Laureate in Physics, 1979),
In reply to Alex Levine (see above) *New York Review of Books,*
Letters to the Editors, T.S. Kuhn's *Non-Revolution:*
An Exchange, February 18, 1999.

[584] Scientitsts on Inevitability of Progress

Note: Steven Weinberg's comments in item **[583]** are relevant to this section.

 Scientists strive for progress through reason with the expectation that logic will win the day—but not, necessarily, right away. Consider these (tongue in cheek?) thoughts of Max Karl Ernst Ludwig Planck.

—JdP

An important scientific innovation rarely makes its way by gradually winning over and converting its opponents. What does happen is that its opponents gradually die out, and that the growing generation is familiarised with the ideas from the beginning.

—Max Planck (1858–1947)
from *Scientific Autobiography and Other Papers,* 1949.

[585]

 *Cheery voices of optimism: Although the question of the inevitability of progress may be debated among scientists and philosophers, there is one mathematician for whom there are no nagging doubts. In item **[443]**, Hermann Hankel tells us that in mathematics, and in mathematics alone, progress is, in fact, ever increasing. Then, in item **[675]**, Martin Gardner takes a Platonist viewpoint with the assertion that mathematicians from anywhere in the Universe will all discover (not invent) the same fundamental ideas.*

Without exception, a mathematical theorem that was true centuries ago, is still intact and true today. The (mostly)

deductive process of mathematics does not seem to require modifications of the type we see in the experimental sciences where the (mostly) inductive, or inferential reasoning process often requires "mid-course" corrections in response to new observations.

—*JdP*

Note: For a discussion of induction and deduction, see items **[160]**–**[171]**. For comments on falsifiability, see item **[676]**.

Progress: The Evolving Value of Carrots

[586]

Change it must. But when language does change, *is it always for the better—a sign of progress?*

For example, the phrase "carrot and stick," started out with one meaning but, in time, took on a totally different one. There are several tellings of the origins of this phrase.

One version, illustrated in **[587]**, *has it that once upon a time, a clever farmer tricked his stubborn donkey into pulling the cart ever forward by using a stick to dangle a tempting carrot before his eyes. Presumably, this teaches us the following:*

Moral: *Resources, such as yummy carrots, can be used both sparingly and effectively.*

[587]

However, today's usage of "carrot or stick" has nothing to do with cleverness or economical use of resources. Current usage conveys the sense of "honey versus vinegar" or "reward versus punishment"—as if carrots have suddenly assumed enormous value, like honey, gold, or beer. For example, in a newspaper article on crowd control for Japanese youths celebrating their 20th birthdays, we read,

> *"Dangling carrots [meaning, beer for the youths] were supplemented with barely sheathed sticks [meaning, riot police on alert]."*

> *—Los Angeles Times, Jan. 15, 2002, pg. A3.*

Returning to our original question: Does this particular evolution in our language represent progress? How can archeologists and sociologists thousands of years from now judge our culture as truly progressive when they see us value cleverness one day and carrots the next?

(Based on a letter by the author read by Liane Hansen on NPR's Weekend Sunday, *21 Mar. 1999.)*

—JdP

[588]   A Required Rebuttal to Galileo   Before publication of Galileo's seminal paper *Dialogo Sopra i due Massimi Sistemi del Mondo* (*Dialogue Concerning the Two Chief World Systems*) was permitted in the journal *Saggi di Naturali Esperienze*, a Church-sanctioned rebuttal had to be published.

It fell to Rev. D. Agostino Calmet to provide the "counterbalance" to Galileo's revolutionary paper, in which he promotes the heliocentric solar system based on his observations of Jupiter, and the phases of Venus.

What could Calmet possibly write? His paper, *Dissertazione Soura il Sistema Del Mondo*, dealt with "la struttura della terra," the structure of the earth, its revolutions and its movements. Calmet called on selected passages from the Hebrew scriptures, along with his considerable rhetorical skills to claim, "the System of the World of the ancient Hebrews was very different from that of our own." This was consistent with his disparaging rebuttal of another unfortunate contemporary who had the temerity to claim the opposite, namely, "Che il mondo di Mose e lo stessissimo, che quello di Cartesio," that the world of Moses is the same as that of Descartes.

It is but a short walk from the Ponte Vecchio to the Ponte Nuovo, but it is as far from Ptolemy to Copernicus, as it is from the "Abjuratio Galileo" of the Priest Calmet, to the "Dialogo" of the scientist Galileo.

—Allan Edelson*, University of California, Davis
The observations above derive from Professor Edelson's own translation of Calmet's Italian text in the journal, *Saggi di Naturali Esperienze*.

[589]

What Should We Know About Ancient Greece?
Following are excerpts from Sandro Graffi's review of Lucio Russo's La Rivoluzione Dimenticata, *which was, published in the* Notices of the Amer. Math. Soc. *(See full attribution at end.) Summarizing headers have been added.*

—JdP

The Forgotten Revolution (La Rivoluzione Dimenticata)   In [his book,] *The Forgotten Revolution* (In Italian, *La Rivoluzione Dimenticata*), Lucio Rosso, a probabilist at the University of Rome ... sets out to reconstruct Hellenistic science between ... 331 B.C. and 145 B.C., the golden age of science in antiquity.

Hellenistic science was first-rate: [The] first conclusion represents an innovation, [namely, that] the Hellenistic scientists were no simple "fore-runners" or "anticipators" of modern science... basically amateurish, unlike the present-day scientists and technologists. On the contrary, they were real pros.

Hellenistic science rediscovered in XVII century: The second conclusion goes even farther: [Just as] the Renaissance was based on the recovery of classical culture, the post-Renaissance scientific revolution of the XVII century was [based on] the conscious recovery of the Hellenistic science.

Inverse square law: Examination of original sources, ... many more than the traditional ones, ... yields new findings in the history of science—One [finding is] the discovery of the inverse square law of gravitation in Hellenistic times, ...

Who owns Euclid's definitions?   ... the other [finding of Russo is] that Euclid's definitions are not in [his] original text. ... Toward the beginning of the second century A.D., Heron of Alexandria found it convenient to introduce definitions of the elementary objects in his commentary on Euclid's *Elements*, which had been written 400 years before. All manuscripts of the *Elements* copied ever since included Heron's definitions without mention, whence their attribution to Euclid himself. The philological evidence leading to this conclusion is quite convincing.

Preserve the worst, abandon the best: Actually, the author's main point about Hellenistic mathematics is its methodolgical nature: Even more important than what Euclid, Archimedes, and Eratosthenes actually discovered, is the method they introduced, namely, the axiomatic, deductive way of argumentation which characterizes mathematics. ... The Hellenistic scientific revolution was forgotten precisely because that scientific method was abandoned in antiquity and its recovery was exceedingly slow. ...

Why did the selection process work in reverse, saving some of the worst and throwing away much of the best? ...

This is a major historical problem, strongly tied to the even bigger one of the decline and fall of antique civilization itself.

—Sandro Graffi, University of Bologna, Italy
From a review of *La Rivoluzione Dimenticata*, by Lucio Russo,
published in *Notices Amer. Math. Soc.*, vol. 45, no. 5, May 1998.

Note: See also item [46] for the roles of Isaac Newton, Edmond Halley and Robert Hooke in the development of the Inverse Square Law.

Will Technology Go Awry?

[590]

Bill Joy, is considered to be a major figure in the history and development of information technology. He was cofounder and Chief Scientist of Sun Microsystems and was co-chair of the presidential commission on the future of IT research. He is also a coauthor of The Java Language Specification. In the following, Joy comments on ethical problems and dangers arising from misjudgment in the use of technology.

—*JdP*

[591]   Bill Joy's Comments Our most powerful 21st-century technologies —robotics, genetic engineering, and nanotech—are threatening to
make humans an endangered species.

From the moment I became involved in the creation of new technologies, their ethical dimensions have concerned me, but it was only in the
autumn of 1998 that I became anxiously aware of how great are the dangers facing us in the 21st century. I can date the onset of my unease to the
day I met Ray Kurzweil, the deservedly famous inventor of the first reading
machine for the blind and many other amazing things.

...I had always felt sentient robots were in the realm of science fiction.
But now, from someone I respected, I was hearing a strong argument that
they were a near-term possibility. I was taken aback, especially given Ray's
proven ability to imagine and create the future.

—Bill Joy from "Why the Future Doesn't Need Us,"
Wired Magazine 8 April 2000,
http://www.wired.com/wired/archive/8.04/joy_pr.html.

[592]   Isaac Asimov's Three Laws of Robotics

1. A robot may not injure a human being, or, through inaction, allow a
   human being to come to harm.

2. A robot must obey the orders given it by human beings, except where
   such orders would conflict with the First Law.

3. A robot must protect its own existence, as long as such protection does
   not conflict with the First or Second Law.

—Isaac Asimov from his book, *I, Robot, 1950.*

[593]   Hans Moravec's Claim: To Be a Robot is Good Robotics guru,
Hans Moravec, who foresees the gradual transformation of human beings
into robotic lifeforms, says Joy's call to relinquish certain technologies is
futile.

"We will turn into robots. It's both inevitable and desirable," he said.
Moravec views this transformation as a natural part of the evolutionary process.

"It's bigger than we are. We are merely components within it." Joy
says it's dangerous to treat technology as a power outside of our control.

"We don't have to make our moral choices subject to Darwinism. That's what makes us human," he said.

> —Tihamer Toth-Fejel from "A Response to Bill Joy's 'Why the future doesn't need us,'" *Nanotechnology Magazine*, May 20, 2000, http://nanoquest.com/nanozine/FutureNeedsBillJoy.htm.

PROOF

Note: See also Kurt Gödel's Incompleteness Theorem, item [**324**], which guarantees that within any reasonable logical system (the Gödel Box), not all true statements are provable.

[594]

> *Like the short-lived butterfly that knows only the warmth of Spring, we often forget that our own warm season of progress came only after many an intellectually cold winter. Read on.*
>
> *—JdP*

Students have difficulty imbibing the spirit of proofs because their need and purpose are not obvious. [To see this, consider] the fact that all the classical civilizations; Egypt, Babylon, India, and China, had a practical geometry, but none treated geometry as a deductive science..., [R]ecipes, more or less correct, were followed and learned for another millennium and a half...before the Greeks devised a logical system that enabled them to demonstrate, on very general assumptions, that the area of a triangle must always equal half the product of its altitude and its base.... Students should not become impatient if they do not immediately understand the point of geometrical argument. Entire civilizations missed the point altogether.

> —J. L. Heilbron, *Geometry Civilized. History, Culture, and Technique*, Clarendon Press, Oxford, 1998, pg. 3.

Note: See also "Is Progress Monotonically Increasing?" item [**580**] and [**443**]. For more on George Boole who first articulated rules of symbolic logic in 1847, see item [**319**].

[595]   We cannot pretend to offer proofs. Proof is an idol before whom the pure mathematician tortures himself. In physics we are generally content to sacrifice before the lesser shrine of Plausibility.

—Sir Arthur Eddington (1882–1944)
The Nature of the Physical World, reprint edition (May 1995)
AMS Press; ISBN: 0404604781.

[596]   A proof tells us where to concentrate our doubts.

—Morris Kline (1908–1992)

[597]

A good proof is one which makes us wiser.

—Yuri I. Manin, Max Planck Institute for Mathematics, Bonn, Germany, and Steklov Mathematical Institute, Moscow, Russia. Recipient, 1994 Frederic Esser Nemmers Prize in Mathematics, *A Digression on Proof, The Two-year College Mathematics Journal*, 12 (2), pp. 104–107, 1981.

[598]

Definition: Reductio ad absurdum (proof by contradiction) is the technique that proves a statement is TRUE by assuming the opposite and then showing that (at least) one unpleasant contradiction will follow.

—JdP

Note: For Hardy's tribute to and acknowledgement of this technique, see item **[603]**.

[599] **Example:**

Objective: Prove
"$\sqrt{2}$, the square root of 2, *cannot* be written as a fraction, m/n, where m and n are postive integers."

1. Assume the opposite:
   $\sqrt{2}$, the square root of 2, *can* be written as a quotient of integers m and n. Specifically,

[600]
$$\sqrt{2} = \frac{m}{n}$$

where integers m and n factor as

$$m = (p_1 p_2 \ldots p_i \ldots p_k), \qquad n = (q_1 q_2 \ldots q_j \ldots q_l)$$

for prime numbers

$$p_i, \text{ where } i = 1, 2, \ldots, k, \quad \text{and}$$
$$q_j, \text{ where } j = 1, 2, \ldots, l.$$

Since any common prime factors in m and n can be cancelled out in the quotient m/n in **[600]**, we may assume that no p_i equals any q_j. The k-term factoring of m and the l-term factoring of n in display **[600]** is unique in each case except for order.

2. If $\sqrt{2} = m/n = (p_1 p_2 \ldots p_k)/(q_1 q_2 \ldots q_l)$, then squaring implies

[601] $2 = m^2/n^2 = (p_1 p_2 \ldots p_k)^2/(q_1 q_2 \ldots q_l)^2, \quad$ which implies

$$2(q_1^2 q_2^2 \ldots q_k^2) = (p_1^2 p_2^2 \ldots p_l^2) \tag{A}$$

2a. One awkward contradiction to Line (A) of **[601]** above is that there is an odd number of unique prime factors on the left-hand side (2 is one of the primes) and an even number of unique prime factors on the right-hand side. Since there are no other prime decompositions for m^2 and n^2, the odd-even mismatch in Line (A) of **[601]** says it cannot be valid.

2b. Another contradiction produced by assuming Line (A) of **[601]** is valid, is that the "2" on the left-hand side forces a "2" to appear on the right-hand side also (both sides of Line (A) of **[601]** should display exactly the same unique prime factors). This means that "2" is a common prime factor to both the numerator m and the denominator n, contradicting our assumption in Step 1 that there are no common factors.

Step 3: Avoid the embarrassing contradictions wrought by 2a and 2b by abandoning the assumed truth of the initial hypothesis that $\sqrt{2} = m/n$. The opposite must therefore be true, namely, that $\sqrt{2}$ *cannot* be represented as a fraction.

—JdP

[602]

Bonus: *The proof above provides a good example of what often happens in mathematics—a proof will often extend to a more general situation with very little modification. In our case, the proof that*

the square root of 2 is never a fraction.

can be extended, almost word-for-word, to prove that

the square root of any prime number p is never a fraction.

Try it and see!

—JdP

[603] *Reductio ad absurdum*, which Euclid loved so much, is one of a mathematician's finest weapons. It is a far finer gambit than any chess play: a chess player may offer the sacrifice of a pawn or even a piece, but a mathematician offers the game.

—Godfrey H. Hardy (1877–1947)
A Mathematician's Apology, London,
Cambridge University Press, 1941.

Note: *Reductio ad absurdum* is defined in item **[598]**.

[604]   We are not very pleased when we are forced to accept a mathematical truth by virtue of a complicated chain of formal conclusions and computations, which we traverse blindly, link by link, feeling our way by touch. We want first an overview of the aim and of the road; we want to understand the idea of the proof, the deeper context.

—Hermann Weyl (1885–1955)
Unterrichtsblätter für Mathematik und Naturwissenschaften, 38,
177–188 (1932). Translation by Abe Shenitzer appeared in
The American Mathematical Monthly, vol. 102, no. 7,
August-September 1995, pg. 646.

[605]

Do we know what a proof is anymore? *The following discussion at Table Seven of the Starlight Café concentrates on the use of probability (usually using computers) to "confirm" whether a number is prime or not. See item* **[532]** *for the definition of prime, items* **[534]** *and* **[535]** *for more on Mersenne primes, and item* **[219]** *for a discussion of the four-color problem and its proof by computer only.*

—JdP

[606]   "I thought I knew what a proof was," Hutch said. "At least before we started to prove a number is prime only with a high probability!" Both Anvil Willie and Cordelia seemed confused by her remark. "Hold on! Didn't you just tell us that a proof is just a sequence of logical steps that you take when you start from basic axioms, postulates, definitions, and theorems?" Cordelia asked.

"Oh yes, I remember," Hutch sighed. "The journey from axioms and definitions to theorems, used to be all there was to a proof. But since digital computers have made concerns for security a reality, 'proving' that very large numbers are prime, for example, has changed the notion of what a proof is."

We listened as Hutch went on to describe her concerns. Paraphrasing, here is what she said:

[607]   In 1640, Pierre de Fermat, who is considered to be the founder of modern number theory, produced a "little theorem," which, by the way, is different from his famous $x^n + y^n = z^n$ theorem, which was proved by Andrew Wiles in 1995.

Fermat's little theorem says if a positive integer $p > 1$ is a prime number (not divisible by any positive integer except 1 and itself as per definition **[532]**), then for any and all integers a, if you divide both a^p and a by p, you will always get the same remainder. In formal congruence terminology,

[608] $$a^p \equiv a \pmod{p}$$

Note: See also Andrew Wiles' comments in item **[182]**.

The "Fermat" test of formula **[608]** above supplies a necessary condition which must hold for all prime numbers p.

Equality (for a pair of numbers a and p) in **[608]** is necessary but not sufficient to flag p as a legitimate prime number. Indeed, there exist numbers that fool the Fermat test **[608]** above. That is,

[609]   **Definition:**   There are numbers p that are not prime and yet have the property that a^p is congruent to $a \pmod{p}$ for all integers a. Such "prime-poseur" numbers p, called **Carmichael numbers**, are extremely rare, so the Fermat test is quite reliable in practice.

Note: See also item **[507]** for a quote from Robert D. Carmichael.

[610]   **Enter Probability:**   Here is how probability determines whether integer p is a prime or not. First, choose an integer a at random. If substitution of candidate p and random integer a into **[608]** fails, then p is not a prime and we are done. But, if equation **[608]** holds true, then p is only probably a prime, since some non-primes will satisfy **[608]** as well.

What to do? Take a second randomly chosen integer a and put it, along with "probable prime" p, into **[608]**. If equation **[608]** fails, then we are done since p is not a prime. But if equation **[608]** holds true, then p is probably a prime—but the probability is increased, since it has now passed the Fermat test twice.

With successive choices of integers a, continue the Fermat testing and the more times integer p passes the test **[608]**, the more likey it is to be an authentic prime. As it turns out, with only a few successes, the likelihood of p being a prime can be very, very high.

"If the likelihood is high that p is a prime number—something like four gazillion out of four gazillion and one," Cordelia offered, "then isn't that good enough?"

Hutch responded by noting that effective encryption is based on the fact that multiplying large numbers is easy, but "un-multiplying," or finding the factors, is difficult. So in the real world, we don't need an air-tight mathematical proof that a number is prime. For effective security in the multiplying of large "prime" numbers, a high probability of prime is good enough.

—John de Pillis from *Starlight Café Conversations:*
An Illustrated Dictionary from Table Seven. (See pg. 319.)

PYTHAGOREAN THEOREM

Preview: With very little physics, item **[626]** *shows how the Pythagorean Theorem is used to explain time dilation and distance contraction in Special Relativity.*

—*JdP*

[611]   Statement of the Pythagorean Theorem: For any triangle with sides having length a, b, and c, we have

$$c^2 = a^2 + b^2$$

if and only if the sides with length a and b form a right angle.

[612]

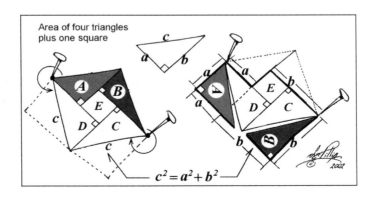

[613] Nonalgebraic proof (one direction: given a right triangle)   We show only one direction of the theorem: IF we have a right triangele THEN. $c^2 = a^2 + b^2$.

Left side of diagram [612]:   With four identical right triangles having sides with length a, b, and c, form the square with four equal sides of length c. The area of the square is c^2.

Right side of diagram [612]:   Now swivel triangle A counter-clockwise, and triangle B clockwise to produce the L-shaped figure on the right which has area $a^2 + b^2$.

Conclusion: Both figures in the diagram, the square and the L-shape, are composed of the same four right triangles and a white square. Their respective areas, c^2 and $a^2 + b^2$, are equal. That is, $c^2 = a^2 + b^2$. Q.E.D.

[614]   The Other Direction of the Pythagorean Theorem

We have only shown one direction of the Pythagorean Theorem, namely,

An a-b-c triangle is a right triangle $\Rightarrow c^2 = a^2 + b^2$.

But the law of cosines shows that the arrow goes the other way, too. That is, the complete, more powerful, if-and-only-if form for the Pythagorean theorem is:

An a-b-c triangle is a right triangle $\Leftrightarrow c^2 = a^2 + b^2$.

—JdP

We now provide four applications which illustrate why the Pythagorean theorem enjoys celebrity status as one of the most important theorems in mathematics:

[615]   APPLICATION 1: The Pizza Connection

(Via Robert Osserman and Tom Lehrer. See item **[616]** following for details.)

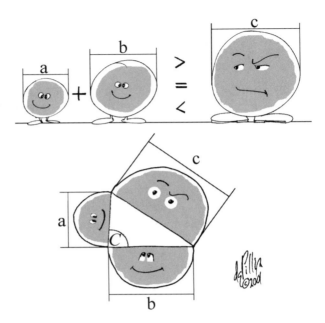

Question: Which gives you more pizza—the small pizza plus the medium pizza, or the large pizza by itself?

Answer: Cut each pizza in half and form the triangle whose sides are the respective diameters a, b, and c shown above. If the angle C is a right angle, then $a^2 + b^2. = c^2$. which means the areas of the small pizza plus the area of the medium pizza is equal to the area of the large pizza.

The large pizza has a larger area than the other two pizzas once the hypotenuse increases—that is, once the angle C is larger than 90 degrees. Similarly, the large pizza has a smaller area than the other two pizzas if the angle C is less than 90 degrees.

[616]   Source: Yes, the pizza version of the Pythagorean Theorem [came to me] when I was preparing for our FermatFest, organized when Wiles announced his proof of Fermat's Last Theorem. ([There is a] video of this event where I demonstrate this.) I should say, though, that the idea was inspired by an earlier version using square birthday cakes that Tom Lehrer had mentioned to me.

—Robert Osserman*, Mathematics Sciences Research Institute
(MSRI), Berkeley (private communication).

Note: See also Tom Lehrer, items **[114]**, **[193]**, **[315]**, **[746]**.

[617]   APPLICATION 2: Breaking the Bonds of Three Dimensions

Find the length of vector (x_1, x_2) in 2-dimensional space: The formula $c^2 = a^2 + b^2$ (item **[611]**) of the Pythagorean Theorem can be expressed directly in terms of Cartesian co-ordinates in 2-dimensional space. As the diagram of item **[618]** shows, hypotenuse c corresponds to length l_2 and the sides of the triangle, a and b, respectively, correspond to lengths of x_1 and x_2. Accordingly,

[618]

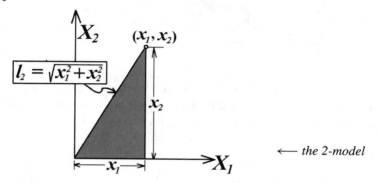

⟵ *the 2-model*

formula $c^2 = a^2 + b^2$ (item **[611]**), becomes $l_2^2 = x_1^2 + x_2^2$ or, after taking the positive square root of each side, we get $l_2 = \sqrt{x_1^2 + x_2^2}$. That is, for two-dimensional vector (x_1, x_2),

[619]   *Length* $[(x_1, x_2)] = \sqrt{x_1^2 + x_2^2}.$ ⟵ *the 2-formula*

[620]

Preview: As we shall see, Formulas of items **[619]** *and* **[622]** *will provide the template that allows us to soar beyond the limits of 3 dimensions. The concept and pattern of the formulas will remain valid for all dimensions, even though pictures or models, such as the triangles of items* **[618]**, *will fail.*

—JdP

Find the length of the vector (x_1, x_2, x_3) in 3-dimensional space: The length l_3 of (x_1, x_2, x_3) is the length of the hypotenuse of the striped right triangle shown in Figure of item **[621]**. This triangle has two perpendicular sides with respective lengths

[621] $l_2 = \sqrt{x_1^2 + x_2^2}$ and x_3.

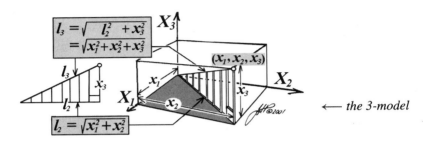

⟵ *the 3-model*

Apply the Pythagorean Formula, item **[611]**, to these lengths $l_2 = \sqrt{x_1^2 + x_2^2}$ and x_3, (illustrated in Figure **[621]**) to find hypotenuse length l_3. That is, for three-dimensional vector (x_1, x_2, x_3), we have

[622]

$$Length\,[(x_1, x_2, x_3)] = \sqrt{l_2^2 + x_3^2} = \sqrt{x_1^2 + x_2^2 + x_3^2}. \quad \longleftarrow \textit{the 3-formula}$$

Find the length of the vector $(x_1, x_2, x_3, \ldots, x_n)$ in n-dimensional space. With dimension $n > 3$, we no longer have intuitive pictures or models like the 2-model of figure **[618]** for (x_1, x_2), or the 3-model of figure **[621]** for (x_1, x_2, x_3).

No matter. We take our imagination *beyond* pictures.

Once we extend the pattern or template of formulas **[619]** and **[622]**, a meaningful definition for the *length* of $(x_1, x_2, x_3, \ldots, x_n)$ emerges. No pictures are necessary.

We borrow the pattern of items **[619]** and **[622]** this way:

[623] $Length\,[(x_1, x_2, x_2, \ldots, x_n)]$

$$= \sqrt{x_1^2 + x_2^2 + x_3^2 + \cdots + x_n^3}. \quad \longleftarrow \textit{the n-formula}$$

With **[623]**, we have extended the Pythagorean Theorem, item **[611]**, and the idea of length to spaces having dimensions higher than three. Our imagination has gone beyond the pictures and models of mere three-dimensional space.

[624]

Nothing in **[623]** *that defines length in n dimensions* contradicts *the intuitive notions of length we are comfortable with in formulas of items* **[619]** *and* **[622]** *where we have pictures to aid us. With* **[623]**, *we have* extended *the notion of "ordinary" length to length of vectors we can no longer visualize.*

—JdP

[625]

So far, we increased dimension, from dimension two to higher ones, in order to extend the idea of length with the Pythagorean Theorem. But we can also take **[623]** *down in dimension—down to dimension one. Using the format of* **[623]**, *we see that for any 1-dimensional vector x_1 on the number line (positive or negative), its length is*

$$Length\ [(x_1)] = \sqrt{x_1^2} = |x_1|$$

—JdP

[626]   APPLICATION 3: Pythagorean Theorem in Special Relativity

Time Dilation of Special Relativity in a Nontechnical Nutshell

[627]

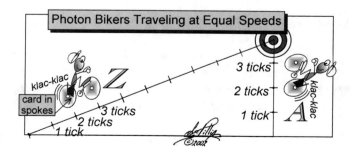

Setting Up Two Identical Clocks: *Diagram* **[627]** *shows two bicycling photons traveling at the same speed. Imagine that each photon places in its rear wheel a baseball card that creates a ticking sound. Since both photons are pedaling at the same speed, the ticking rates are the same—in this way, the bicycling photons produce identical ticking clocks.*

The Interpretation: *The diagonal photon takes more ticks (time) to reach the bull's-eye than does the vertical photon. To say it another way, until the photons reach the bull's-eye, observer Z of the diagonal photon ages more than observer A of the vertical photon.*

What Special Relativity Adds: *The two photons of Diagram* **[627]**, *are, in fact, two views of the same photon! The diagonal mode is observed by Z (for Zelda) and the vertical mode is observed by A (for Abbey). Since there is only one photon, it (they) hit the bull's-eye at the same moment. To see how two photons of* **[627]** *are viewed as one photon, see Diagram* **[631]**. *(Note that the bull's-eye of Diagram* **[627]** *becomes the railroad car ceiling of Diagram* **[631]**.)

Why Time Is Elastic: *While observing the photon traveling to the bull's-eye, Z counts more ticks than A does. (Remember, each observer sees the single photon bicycling at the same speed—this is Property (a) of* **[628]**—*so each observer counts ticks from identical clocks. That is, more time has passed for Z than for A to observe the same event. Z ages more (or faster) than A does.*

—*JdP*

Two Assumptions

In this section, using the core idea of Diagram **[627]**, *the Pythagorean Theorem, and a minimal amount of physics, we will see that Einstein was correct: An observer of a moving object (a train, for example) will see*

- *the train's clocks run slower, and*

- *the train's length, in the direction of motion, will shrink.*

The "minimal" physics we need is expressed in just these two properties, namely, **[628]** *and* **[629]**.

Assumptions

[628] Property (a): In a vacuum, the speed of light, c, is perceived to be the same for all observers, regardless of the motion of the observers. (The speed of light in a vacuum for all observers is 186,000 miles/sec, or, not to put too fine a point on it, 299, 792.458 . . . kilometers/sec.) Unlike chasing a train to make it appear to go slower, you can not chase a beam of light to make it appear slower. Whether you run or stay put, the speed of the light will be the same. This is a counter-intuitive idea!

You can chase a ball to make its speed seem slower to you.

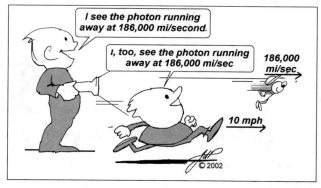

You *cannot* chase light to make its speed seem slower to you.

[629] Property (b): For any two observers A and B, each imagines herself to be stationary while the other observer is perceived to be the one in motion. The perceived speed of one observer **relative** to another is the **same** for each observer.

Time dilation as a consequence of Pythagorean Theorem.

[630] Since the speed of light is the same for all observers, (**Property (a)**, item **[628]**), it follows from the Pythagorean Theorem that time itself expands and contracts, depending on the observer and the motion of the clock being observed.

Let's see how this works.

Consider Diagram **[631]**, (an expansion of Diagram **[627]**) in which a railroad car with floor-to-ceiling height d_i is traveling at v feet per second along a horizontal track. (Abbey and Zelda see each other as moving at v feet per second.)

[631]

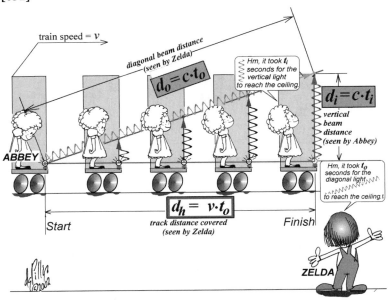

Use Pythagorean Theorem to see how t_i compares with t_0.

Here is what happens when both girls' clocks are synchronized with each other to read zero exactly when a light beam starts its journey from the floor toward the ceiling.

1. **The (vertical) light beam from Abbey's inside point of view:** When the vertical light, with speed c, travels distance d_i from the floor and strikes the ceiling, Abbey reads her (inside) clock at t_i seconds. The vertical inside height, d_i, of Abbey's railroad car is obtained by combining quantities c, d_i, and t_i according to the formula

$$(\text{distance} = \text{velocity} \times \text{time})$$

   to obtain

[632] $d_i = c \times t_i$

2. **The (diagonal) light beam from Zelda's outside point of view:** The *same* light beam, which travels vertically from Abbey's point of view, is seen as traveling diagonally from Zelda's point of view. When the light, with speed c, travels the (diagonal) distance d_0 from the floor, and strikes the ceiling, Zelda reads her (outside) clock at t_0 seconds. The diagonal distance, d_0, of the observed light path is obtained by combining the quantities c, d_0, and t_0 in the usual way: distance = velocity × time. This yields

[633] $d_0 = c \times t_0$

3. **Horizontal distance, or length of track, traveled by the car from Zelda's point of view:** The horizontal distance, d_h, that Zelda sees covered by the railroad car, going at speed v, is

[634] $d_h = v \times t_0$

Zelda says Abbey's clock runs slower: From equations in items [632] and [633], we see that $d_i < d_0$ which implies the light beam's "diagonal time," t_0, outside the railroad car, is greater than the "vertical time," t_i, inside the car. This means that Zelda sees Abbey's clock running slower than her own clock. (Outside the railroad car, Zelda sees Abbey's clock, inside the railroad car, tick fewer times than her own clock.) Therefore, Zelda sees herself aging faster than Abbey, who is in motion.

—JdP

The note above is a qualitative remark. But the Pythagorean Theorem (item **[611]**) tells us, quantitatively, exactly how much slower Abbey's moving clock runs relative to Zelda's stationary clock.

Since items **[632]**, **[633]**, and **[634]** represent the lengths of the three sides of the right triangle diagrammed in item **[631]**, the Pythagorean Theorem yields

$$c^2 t_i^2 + v^2 t_0^2 = c^2 t_0^2.$$

After some algebraic manipulation, the equation above gives us the relationship between Abbey's time t_i inside the moving car and Zelda's time t_0 outside the moving car as follows:

[635] $$\frac{t_i}{t_0} = \sqrt{1 - \frac{v^2}{c^2}} \quad \text{or} \quad t_i = t_0\sqrt{1 - \frac{v^2}{c^2}} = t_0 \cdot \alpha(v),$$

which defines the relativity "reduction factor,"

[636] $$\alpha(v) = \sqrt{1 - \frac{v^2}{c^2}} < 1 \quad \text{if} \quad v > 0.$$

Summary: From any observer's point of view, all clocks moving at speed v run slower by the relativity reduction factor of $\alpha(v) < 1$ given by **[636]**. For example, from Zelda's stationary point of view, Abbey's moving clock, moving at speed v, runs slower by this same factor $\alpha(v) < 1$.

[637]

Mathematics says matter can't go faster than light. *Since a square root of a negative number is not a positive number, it is the mathematics of item* **[636]** *that forces matter to have speeds v less than c, the speed of light. That is, item* **[636]** *says we must avoid $1 - (v/c)^2$ becoming negative, which happens only when v is greater than the speed of light c.*

—JdP

[638]

We touch on the question of symmetry of the relativistic argument. Reverse the situation and take Abbey's point of view (when she sees herself as stationary). Now Zelda's

clock is the clock that is moving. Hence, Abbey sees
Zelda's clock running slower than her own clock.
However, should Abbey's train suddenly reverse course
and return, then symmetry is lost. This is because Abbey
would be changing her "inertial frame of reference" while
Zelda would retain her original inertial frame of
reference.

—JdP

[639]   Distance contraction of moving objects Moving objects have
lengths that are observed to shrink in the direction of their travel. Why?

To see this, let us decide that the (moving) length we will measure is
the length of the outside track covered by Abbey's railroad car during the
time the beam travels from floor to ceiling. The track is moving relative
to Abbey, who is inside the car, and is stationary relative to Zelda, who
is outside the car. Abbey and Zelda will perceive different lengths of the
track.

Remember—distance will not be measured by a tape measure, but by
the relationship

[640]   distance = velocity × time.

Although both Abbey and Zelda share the same perceived velocity v, (see
item **[629]**), we have seen in item **[635]** that their respective times for the
floor-to-ceiling journey of light will have different values, namely, t_0 for
Zelda and t_i for Abbey. (The two floor-to-ceiling distances are different,
after all.) It must follow from item **[640]** that the horizontal distances of the
track, as measured by Zelda and Abbey, respectively, are

[641] d_Z (stationary track length measured by Zelda) $= v \times t_0$

             d_A (moving track length measured by Abbey) $= v \times t_i$

where

$$t_i = \alpha(v) \cdot t_0, \text{ and } \alpha(v) = \sqrt{1 - (v/c)^2}, \quad \text{from item } [635].$$

Substituting this expression for t_i into item **[641]**, gives us the track length
covered (during the floor-to-ceiling journey of light) as

[642]
$$d_A = d_Z \times \sqrt{1 - (v/c)^2}$$

d_A = shorter track length seen by moving Abbey.	d_Z = longer track length seen by stationary Zelda.

Moving tracks shrink from their stationary length d_Z to their "moving" length d_A. Abbey and Zelda see the same track. But when the track is in motion (as Abbey sees it from inside the moving car), it is perceived to be shorter than when it is stationary (as Zelda sees it outside the moving car).

[643] Application 4: Michelson-Morley's Valuable Failure

Today, Hutch brought a guest, Lisette, to Table Seven of the Starlight Café. Our guest was knowledgeable in physics and, more importantly, was willing to share her insights. We ordered up coffee and the special tray of snacks, for this, indeed, was a special occasion.

After the introductions, Lisette said, "If I understand what Hutch has told me about your discussions, then I think you might like to hear about the Michelson-Morley experiment."

Michelson-Morley? Why, we wondered aloud, should this particlular experiment be of interest to us? The answer, according to Lisette was that this important experiment combined many particularly interesting features.

*First, the Michelson-Morley experiment makes a good story because it had a surprise ending. The experimenters set out to prove one thing—the existence of an "ether," a light-transmitting medium whose "currents" affect the speed and direction of light. But they proved the opposite, namely that there is no such ether. This was an unexpected surprise. (Einstein, who claimed to be unaware of the Michelson-Morley experiment, assumed in his Special Relativity theory, item **[628]**, the constancy of the speed of light. He did not rely on an assumed ether.)*

Secondly, Michelson-Morley harmoniously brings into play, elements of theory, logic, and experimentation, namely,

- **The Pythagorean Theorem:** *used to predict (deduce) the quantitative effects of a cross-current of ether on a beam of light traveling perpendicular to it (Diagram* **[645]**),

- **Logic:** *The search for a cause (inference) to explain the constancy of the speed of light in Michelson-Morley's "failure" (item* **[657]**). *The unavoidable Four-Steps of logic, using the contrapositive and modus ponens in physics (see definition* **[659]**),

- **Theory:** *How Maxwell's partial differential equations correctly predicted (deduced) the precise speed of electromagnetic waves (including light) in a vacuum. (items* **[661]**, **[662]**).

Here follows a summary of Lisette's comments.

—JdP

[644]   Assumptions: Henceforth, and through item **[654]**, we will assume:

- The ether exists throughout the Universe as the transmitting medium for light. The speed and direction of the light are affected by currents and cross-currents of the ether, just as the speed and direction of a motor boat is affected by currents of a river.

In vector terms: The velocity vector of the ether "river" adds to the initial velocity vector of the light-beam photons. The vector sum is the final velocity vector for the photons of light.

- The Earth is assumed to be moving through the ether horizontally, from left to right. Our diagrams will always take the Earth's point of view so that the Earth is seen as stationary and the ether is seen as moving horizontally from right to left. (See Diagrams **[645]**, **[654]**.)

[645]

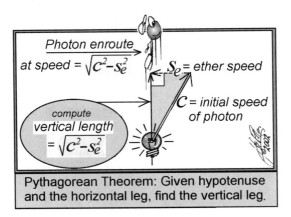

Photon enroute at speed $= \sqrt{c^2 - s_e^2}$

$s_e =$ ether speed

compute vertical length $= \sqrt{c^2 - s_e^2}$

$c =$ initial speed of photon

Pythagorean Theorem: Given hypotenuse and the horizontal leg, find the vertical leg.

[646] *Ether's Effect on a Vertical Beam of Light (A Deduction)* Diagram [645] shows how, relative to a stationary Earth, an almost vertical light beam is pushed onto a vertical path by a horizontal flow of the (assumed) ether.

(a) The photon starts out from the light source, aimed a little to the right of the vertical. Along this path, its emission speed is c.

(b) The ether's horizontal flow (relative to a stationary Earth) pushes the photon leftward onto a true vertical path. The horizontal, left-to-right speed of the deflecting ether is s_e.

(c) The photon, having been pushed onto a vertical path, now has a final (altered) direction and speed.

[647]   vertical speed $= \sqrt{c^2 - s_e^2}$.

This speed is obtained by the Pythagorean Theorem, which says the length of the vertical side of the right triangle in Diagram [645] has the value given by [647].

[648] *Ether's Effect on a Horizontal Beam of Light (A Deduction)* If, relative to an apparently stationary Earth, the ether is traveling horizontally at speed s_e, then a horizontal beam of light will keep its direction (in contrast to the vertical beam in Diagram [645]) but its speed will increase or

decrease, depending on whether the beam travels with or against the ether current. In sum, as observed from the Earth, the beam of light, starting at speed c, has

[649]   horizontal speed $=$ $\begin{array}{l} c + s_e \quad \text{traveling with the ether, or} \\ c - s_e \quad \text{traveling against the ether} \end{array}$

Comparing Speeds of Horizontal and Vertical Beams (A Deduction)
Given the vertical speed **[647]** and horizontal speeds **[649]**, we can compute the time that each beam needs for a round trip (RT) over a common distance D. Use the formula,

[650]   RT time $= D/\text{speed}$

to show that the ratio of round-trip times computes to be:

[651] $\dfrac{\text{RT vertical time}}{\text{RT horizontal time}} = \sqrt{1 - (s_e/c)^2} \left.\begin{array}{l} \\ \\ \end{array}\right\} \begin{array}{ll} < 1 & \text{if speed } s_e > 0, \\ = 1 & \text{if speed } s_e = 0. \end{array}$

[652]

Note: If an ether exists with a measurable drag-inducing speed $s_e > 0$, then, as **[651]** indicates, the vertical round-trip time divided by the horizontal round-trip time should always be less than one. That is, if $s_e > 0$, the horizontal round trip always takes more time than the vertical round trip.

Note: Compare ratio **[651]** with the reduction factor **[636]** from Special Relativity.

[653]   The Michelson-Morley Experiment: In 1887, an ingenious experiment was devised by Albert Michelson and Edward Morley, in order to confirm the predicted (deduced) ratio **[651]** of (vertical/horizontal) round-trip times.

As Diagram **[654]** indicates, an interferometer (invented and built by Albert Michelson himself) splits a light beam into two, in-phase beams. One beam, the horizontal one, travels in the direction of the ether (relative to the Earth's motion), and the other beam, the vertical one, travels perpendicular to the ether.

The separate beams are returned by mirrors; each mirror is placed the same distance, D, from the source.

Now if an ether exists with horizontal speed s_e (relative to the Earth), then the speeds of the vertical and horizontal beams must differ, as described in items **[647]** and **[649]**. That is, the two beams that started out in phase with each other, will return to the source out of phase. This implies that Michelson's interferometer will necessarily show a wave pattern interference caused by the two beams being out of phase on their return.

[654]

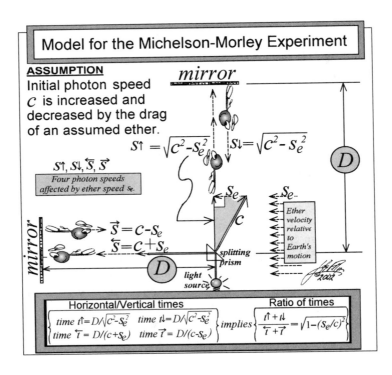

[655] Note: Although we stated that the light beam is split into two mutually **perpendicular** beams, Diagram **[654]** shows the beams (double arrows) starting out by being aimed a little more than 90 degrees from each other. This is necessary to allow the ether to "push" the vertical beam (leftward) back onto a true 90-degree path relative to the horizontal beam.

[656]   The Observed Result? (The ratio that never was.) No wave pattern interference was ever observed in the Michelson-Morley experiment! The waves of both beams were always in phase with each other from start to finish. Time after time, no matter which direction the 90-degree beams were rotated or pointed, there was never any detectable difference in speed between them.

Experiment forces us to acknowledge that ratio **[651]** is exactly equal to one. Mathematically, this implies that, relative to an ever-orbiting Earth, the ether speed, s_e, is always zero! Or there is no ether.

[657] *What Does it All Mean? (Searching for an Inference)*

Question: If there really is an ether that permeates the Universe, why should s_e, its speed relative to the rotating Earth, always be equal to zero?

Possible Answer: The Earth somehow gets stuck to the ether, dragging it all over the Universe.

Question: What could be a cause for (what can we infer from) the speed of light being fixed even when observed from a rotating Earth?

Possible Answer: There could be some (as yet unexplained) shrinkage of distance in the horizontal direction, allowing the horizontal beam to take less time than predicted. (See **[651]** and note **[652]**.) In fact, this was the suggestion of the Irish physicist, George F. Fitzerald (1851–1901) who, in 1892 suggested that distances do, somehow, contract with speed. As it turned out, Fitzgerald was correct (See Special Relativity items **[639]**–**[642]**). However, at the time of his observation, there was no solid theoretical underpinning for this distance-shrinking concept. (See item **[677]** for a relevant limerick.)

These questions, and the grasping, almost desperate answers above, left physcists puzzled for some twenty five years before the ideas of Special Relativity were to emerge. (Special Relativity has no need for a pervasive ether. See items **[628]** and **[629]** for assumptions and items **[630]**–**[642]** for resulting (deduced) phenomena of time dilation and distance shrinking in the theory of ether-free Special Relativity.)

[658] *Logic Meets Physics*
The "failed" experiment of Michelson-Morley not only established the persistent constancy of the speed of light, but it disproved the existence of the ether. Logic allowed for no other outcome.

Here is a Four-Step outline of how the logic of this "dis-proof" works:

Step 1. **(The Michelson-Morley Implication)** The assumptions of **[644]** led to the calculated vertical speeds of **[647]** and the horizontal speeds of **[649]**. From these speeds, we obtained the mathematically valid quotient **[651]**. Ratio **[651]**, in turn, guarantees that the implication,

| P (Ether exists) | implies | Q (Quotient **[651]** is less than 1.) |

is TRUE.

Step 2. **(The Equivalent Contrapositive)** According to item **[30]**, the implication of Step 1. is equivalent to its (equally TRUE) contrapositive, namely,

| $\sim Q$ (Quotient **[651]** is *not* less than 1.) | implies | $\sim P$ (Ether does *not* exist.) |

Step 3. **(Truth of $\sim Q$ by Experimental Observation)** Michelson-Morley's experiment showed, time after time, that the statement,

$$\sim Q = \text{(Quotient } \textbf{[651]} \text{ is } not \text{ less than 1)},$$

is TRUE.

Step 4. **(Deducing the Non-existence of Ether)** Combining Step 2 and Step 3): Since implication,

$$(\sim Q \text{ implies } \sim P) \text{ is TRUE,} \qquad \text{(from Step 2)}$$

and the premise, (hypothesis)

$$(\sim Q) \text{ is TRUE,} \qquad \text{(from Step 3)}$$

it follows that the deduced conclusion (consequence)

$$(\sim P)$$

is necessarily TRUE as well.

Finally, to say "$(\sim P)$ is TRUE," is to say "Ether does not exist." (Step 2 provides the definition of $(\sim P)$.)

Note: The chain of reasoning which we used in Step 4, formally known as *modus ponens*, is defined as follows:

[659]   **Definition:** *Modus Ponens*   is the three-step technique that proceeds as follows:

(a) If you can show that the implication, $P \to Q$ is TRUE, and

(b) if you can show that the hypothesis, P is TRUE then

(c) Statement Q is TRUE.

If we simply rewrite these three steps with $P \to Q$ replaced by its logical equivalent, $\sim Q \to \sim P$, the contrapositive (see item [30]), then we obtain:

Definition: *Modus Ponens* **(Proof by Contradiction form)**

(a') If you can show that the implication, $\sim Q \to \sim P$ is TRUE, and

(b') if you can show that the hypothesis, $\sim Q$ is TRUE, then

(c') Statement $\sim P$ is TRUE.

[660]

Modus Ponens, Popper and Proof by Contradiction are related. Notice that Step 1 and Step 2 of **[658]** *correspond exactly to replacing line (a) in* **[659]** *with line (a'). Hence, modus ponens is related to proof by contradiction (see definiton* **[598]***).*

Popper's falsifiability of item **[171]** *is a physicist's version of proof by contradiction. This equivalence is illustrated in the four steps of* **[658]** *used to interpret the Michelson Morley experiment. Step 4 finally nails down the fact that a transmitting ether that alters the speed of light can not exist.*

Michelson-Morley experiment provides but one example that shows how logic (modus ponens) *seems to be "hard-wired" into our brains. Such often subconscious reasoning seems to be intrinsic to the anatomy of thought we all share.*

—JdP

Note: The four steps of **[658]** provide an example of Karl Popper's falsifiabliltiy in physics, item **[171]**. Also, David Finkelstein, in item **[676]**, asserts that physical law is constructed only by violating it. See items **[598]** and **[599]** for more on proof by contradiction.

[661] **Theory (Maxwell's Equations):** Startling and confounding as the Michelson-Morley experiment was (showing the speed of light to be stubbornly constant and resistant to the effects of the ether's currents), there already existed a solid theory to explain light's mysteriously fixed speed. Well before the 1887 experiment of Michelson-Morley, James Clerk Maxwell (1831–1879) produced four partial differential equations that, together, form a complete description of the production and interrelation of electric and magnetic fields.

From Maxwell's theoretical equations, the speed of any electromagnetic wave in a vacuum is predicted to be (deduced as) c, exactly the known value of the speed of light in a vacuum.

Merely a coincidence?! (See item **[72]** for Shermer's Baloney Detection Rule no. 6 concerning the value of many sources and observations leading to a single conclusion.) Is light, therefore, an electromagnetic phenomenon? During Maxwell's time, the answer was not known although he certainly had his suspicions. Around 1862, in London, he wrote:

[662]

> We can scarcely avoid the conclusion that light consists in the transverse undulations of the same medium which is the cause of electric and magnetic phenomena.

—James Clerk Maxwell (1831–1879) from
The Life of James Clerk Maxwell, by L. Campbell and W. Garnett,
London, 1882.

[663] **Light Is Electromagnetic After All**   Maxwell's equations described how oscillating electrical charges produced magnetic fields (and vice versa). In 1888, Heinrich Rudolph Hertz (1857–1894) experimentally showed that oscillating electrical charges could also produce electromagnetic (radio) waves, thereby extending the range of phenomena described by Maxwell's equations.

If light, like magnetic and radio waves, is to be electromagnetic and described by Maxwell's equations, then what are the electrical oscillations that generate it?

In 1895, the Dutch physicist Hendrik Antoon Lorentz (1853–1928) postulated (inferred) that oscillating electrical charges within atoms themselves are the source (cause) of light.

In 1896, Lorentz's student, Dutch physicist, Pieter Zeeman (1865–1943) tested Lorentz's idea. Zeeman placed a light source under the influence of a strong magnetic field. Magnetic energy was imparted to the hapless electrons of the source and when the excited electrons calmed down, they released that energy in the form of photons (packets) of light! This spray of photons produced changes in the spectral lines, relative to the spectral lines of the unmagnetized, unexcited light.

These observations confirmed the links between magnetism, electrical charges of the atoms and light. Yes, light is electromagnetic in nature.

Our discussions with Lisette ended later than expected. Unnoticed, a clear night sky appeared through the picture window of the Starlight Café. We could see the stars in profusion, stars whose lights never seemed more beautiful than on this night.

We were feeling a warm calmness after much talk of the Pythagorean Theorem, Michelson-Morley, the nature of light and the mechanics of reasoning. Sometimes, striving and reaching for a level of understanding can be both elevating and tiring.

It was time. We pushed our chairs from Table Seven, stood up and put on our coats to leave.

—John de Pillis from *Starlight Café Conversations: An Illustrated Dictionary from Table Seven.* (See pg. 319.)

Other Views on the Speed of Light

[664]

Those Who Would Freeze Light: *If light passes through any medium other than a vacuum (like water or glass, say), its speed will diminish slightly. At the Rowland Institute for Science in Cambridge, Mass., experimenters are trying to change the adverb "slightly" to "drastically."*

At the time of this writing at least, the hopes and dreams of the Rowland Institute are unproved and highly speculative (See item [214] for Martin Gardner's thoughts on theories without experimental validation.) The Institute seeks to apply their "frozen light" ideas to new optical communications technologies, tabletop black holes, and the development of quantum computers. See the article "Frozen Light" by Lene Vestegaard Hau, Scientific American, July 2001, vol. 285, no. 1, pg. 66.

—JdP

[665]

The Speed of Light Depends on Culture? Some
post-modernists would say that truth (which includes the
laws of physics) is not objective and valid for all people.
Rather, truth is perceived differently by different cultures,
each of which has a right to its particular, equally valid
perceptions. For more on belief in the absence of universal
truth or standards, see item **[105]**. To see how the lack of
standards was revealed by physicist Alan Sokal, see item
[91].

—JdP

Q

Q.E.D. MADE EASY

Here are some time-saving proofs. The methods are quick and they give the right results—mostly. The results are pretty close to ones you find in math books.

[666] Proof by Linearity. This method uses the fact, well-known to students, that all functions are linear.

Example 1.

$$\lim_{x \to 0} \frac{\sin 3x}{\sin 5x} = \lim_{x \to 0} \frac{3 \sin x}{5 \sin x} = \frac{3}{5} \qquad \text{Q.E.D.}$$

Example 2.

Theorem. *If p is a prime $(p-1)! + 1$ is divisible by p.*

Proof. Clearly if p is a prime $(p-1) + 1 = p$.

Apply the operator, !, to both sides to obtain

$$(p-1)! + 1! = p! \qquad \text{or}$$
$$(p-1)! + 1 = p!.$$

Since the right-hand side of the last equation is divisible by p, so is the left-hand side.   Q.E.D.

[667] Proof by Disguised Assumption.   In this method one simply assumes, in some ingenious and time-saving manner, what was to be proved. For example we prove the

Theorem. *There exist an infinite number of primes.*

Proof. Let $p_n = $ the nth prime. Consider the series $\sum 1/p_n$.

If the series diverges, obviously there exists an infinite number of primes.

If the series converges, then $1/p_n \to 0$, or $p_n \to \infty$. Once again, there must exist an infinite number of primes.   Q.E.D.

[668] Proof by Rotation.   It is well known that if one rotates a figure $90°$ counterclockwise and follows it with a $90°$ clockwise rotation, no harm is done. However, the application of this method to algebraic problems is still in its infancy. We illustrate by proving:

Theorem. *If $\frac{1}{\infty} = 0$, then it must follow that $\frac{1}{0} = \infty$.*

Proof. We are given that $\frac{1}{\infty} = 0$. Rotate each side of the equation $90°$ counterclockwise to obtain

$$-|8 = 0$$

Now add 8 to both sides of the equation to get

$$-|0 = 8$$

Now rotate each side $90°$ clockwise to obtain $\frac{1}{0} = \infty$.   Q.E.D.

[669] Proof by Cancellation.   Once legitimate cancellation is mastered, it is tempting to carry the method to "similar" but invalid situations. For example, to show that

$$\lim_{x \to \infty} \frac{\sin 3x}{\sin 5x} = \frac{3}{5}$$

simply cancel sin and x from numerator and denominator.   Q.E.D.

[670] **Formal Methods.**   This class of methods is favored by many engineers and scientists. Just forget the meaning of symbols or the conditions under which equations or operations are valid.

For example to prove that every matrix satisfies its characteristic equation (The Cayley-Hamilton Theorem), do the following: Let

$$c(\lambda) = \det(\lambda I - A)$$

where A is a square matrix. Since any polynomial identity remains valid when the variable is replaced by a matrix, we have

$$c(A) = \det(AI - A) = \det(A - A) = \det(0) = 0.$$

Q.E.D.

[671] **Miscellaneous Methods.**   These methods are probably already familiar. For example, we have

(a) *reductio ad nauseum*, (I said it thirty times already. Aren't you convinced yet?)

(b) proof by handwaving,

(c) proof by intimidation, (Only a silly sausage would disagree with me.)

(d) proof by referral to non-existent authorities, (The Great Sausage says so!)

(e) the method of least astonishment, (Obvious, you see.)

(f) the method of deferral until later in the course,

(g) proof by reduction to a sequence of unrelated lemmas (sometimes called the method of convergent irrelevancies) and finally, that old standby,

(h) proof by assignment (uhh... you do it, OK?).

[672]   I am afraid that I rather give myself away when I explain. Results without causes are much more impressive.

—Sir Arthur Conan Doyle (1859–1930)
spoken by Sherlock Holmes in *The Stockbroker's Clerk*.

R

REALITY-PLATONISM

[673]

Definition of Platonism Plato expands upon the work of Socrates by teaching that the objectively real ideas are the foundation and justification of scientific knowledge. Reality is "something out there," and objectively external. With respect to mathematics, objects exist independent of the human mind, and mathematical statements are objectively either true or false.

A platonist believes that mathematics is discovered rather than invented or created.

—JdP

[674] I doubt that many mathematicians believe in Platonism. It does seem to be true that many mathematicians believe that many mathematicians believe in Platonism; but this isn't transitive.

—Chandler Davis*, University of Toronto

[675] I believe that if mathematicians on any other planet, anywhere in the universe, have a sufficiently advanced knowledge of arithmetic and geometry, they will know the Pythagorean theorem, that pi is 3.14+, and that 113

is prime. Of course, they will express these truths in their own language and symbols. Within formal systems, mathematical theorems, unlike a culture's folkways and mores, and even its laws of science, are absolutely certain and eternal.

—Martin Gardner*, science writer

Note: See item **[430]** for Bertrand Russell's implied Platonism.

[676]

Dear John,

Seeing you after all these years reminded me of our first meeting, the first day of the first class I ever taught. My opening was:

"Physics is the search for the laws of nature."

Next year the *Einheit* of Einstein dawned on me like a conversion experience: The Law is One.

"Physics is the search for the Law of Nature."

Then operationalism hit me and "Law of Nature" became an oxymoron. People make laws, Nature just acts naturally. The law of a system relates our input and outtake operations on the system. So:

- We construct the law of a system only by violating it. Now I begin teaching each term with:

- Physics is an evolving adaptive strategy. The joy of physics is part of the grand strategy.

My philosophy nowadays is not so much against Plato as for Bacon, Newton, Peirce, Einstein, Heisenberg, the Dalai Lama, Smolin, and many other practitioners of the experimental philosophy. Borrowing a word from Whitehead: I'm not against reality, I'm for actuality.

Regards,
David

—David Ritz Finkelstein*, Professor of Physics, Georgia Institute of Technology in a private communication from the author of *Quantum Relativity: A Synthesis of the Ideas of Einstein and Heisenberg*, Springer-Verlag New York, March 1996.

RELATIVITY

Note: See also item [613] for a geometric, non-algebraic proof of the Pythagorean Theorem, and items [626]–[642] for the role of the Pythagorean Theorem in time dilation and distance contraction of Special Releativity.

[677]
> There was a young fellow named Fisk
> Whose fencing was agile and brisk.
>> So fast was his action
>> The Fitzgerald contraction
> Diminished his sword to a disk.

Note: See item [657] for a discussion of the Fitzgerald contraction.

[678] **Physicists at the River** A physicist was staring dumfounded at a rushing river blocking her path. As she wondered how to cross, she saw another physicist on the other side. She yelled "Hey, can you help me get to the other side?" The other physicist replied "You ARE (relatively speaking) on the other side!"

[679]

There was a young lady named Bright,
Who travelled much faster than light,
  She started one day
  In the relative way
And returned the previous night.

—Arthur H. Reginald Buller (1874–1944), British poet
The Lure of the Limerick: An Uninhibited History, by
William S. Baring Gould, New York: Clarkson N. Potter, 1967
(out of print)

[680]

To her friends said the Bright one in chatter,
"I have learned something new about matter:
  As my speed was so great
  Much increased was my weight,
Yet I failed to become any fatter."

—Arthur H. Reginald Buller (1874–1944), British poet
The Lure of the Limerick: An Uninhibited History, by
William S. Baring Gould, New York: Clarkson N. Potter, 1967
(out of print)

S

SCIENCE

[681]   One can measure the importance of a scientific work by the number of earlier publications rendered superfluous by it.

—David Hilbert (1862–1943)
H. Eves, *Mathematical Circles Revisited*,
Boston: Prindle, Weber and Schmidt,1971.

[682]   And then there is the story about a civilization so advanced that a prize was awarded to the mathematician who first realized that the Riemann Hypothesis actually needed a proof!

—Ron Graham*, University of California, San Diego

[683]   Mathematical inquiry, like human cruelty, needs no reason—it merely requires opportunity.

—Emmeline de Pillis, University of Hawaii, Hilo

[684]   A science is said to be useful if its development tends to accentuate the existing inequalities in the distribution of wealth, or more directly promotes the destruction of human life.

—Godfrey H. Hardy (1877–1947)
A Mathematician's Apology,
London: Cambridge University Press, 1941.

Note: As item **[416]** indicates, Hardy took pride in the fact that his pure mathematics, at least, would never be perverted by usefulness. This is one of the most often quoted mis-predictions by a mathematician. Hardy's work found its way into many applications.

[685]   In modern times, the belief that the ultimate explanation of all things was to be found in Newtonian mechanics was an adumbration of the truth that all science, as it grows towards perfection, becomes mathematical in its ideas.

—Alfred North Whitehead (1861–1947)

[686]   Science is the great antidote to the poison of enthusiasm and superstition.

—Adam Smith (1723–1790)
The Wealth of Nations

[687]   Science has become too complex to affirm the existence of universal truths, but it strives for nothing else.

—Henry James (1843–1916)

Note: For more on whether there are absolute truths, see items **[105]** and **[174]**.

[688] Religion is always right. Religion protects us against that great problem which we all must face. Science is always wrong; it is the very artifice of men. Science can never solve one problem without raising ten more problems.

—George Bernard Shaw (1856–1950)
From an address delivered at the dinner for Professor Einstein in
London, October 27, 1930.

[689]

True science is restrictively the study of useless things. For the useful things will get studied without the aid of scientists. To employ rare minds on such work is like running a steam engine by burning diamonds.

—C. S. Peirce (1839–1914)

[690] I have little patience with scientists who take a board of wood, look for its thinnest part, and drill a great number of holes where the drilling is easy.

—Albert Einstein (1879–1955)

[691]   In the space of one hundred and seventy-six years, the Lower Mississippi has shortened itself two hundred and forty-two miles. That is an average of a trifle over one mile and a third per year.

Therefore any calm person, who is not blind or idiotic, can see that in the old Oolitic period, just a million years ago next November, the Lower Mississippi River was upward of one million three hundred thousand miles long, and stuck out over the Gulf of Mexico like a fishing rod. And, by the same token, any person can see that seven hundred and forty-two years from now the Lower Mississippi will be only a mile and three quarters long, and Cairo and New Orleans will have joined their streets together, and be plodding along under a single mayor and a mutual board of aldermen. There is something fascinating about science. One gets such wholesale returns of conjecture out of such trifling investment of fact.

—Mark Twain (1835–1910), born Samuel L. Clemens
Life on the Mississippi

[692]

Most problems in scientific computing eventually lead to solving a matrix equation.

—Gene Golub*, Fletcher Jones Professor of
Computer Science, Stanford University

SETS

[693]

Bertrand Russell was most mortified
When a box was washed up by the tide,
    For he said with regrets,
     "Why, the set of all sets
Which belong to themselves is inside."

—Paul Ritger

[694]

    Am I being stubborn and lazy
    To question a fact that is hazy?
      To say that the empty set
      Lives inside every set
    Is simply, and plainly just crazy.

—John de Pillis

[695]   To be read aloud

Student: Is Juan in the empty set?

Professor: No Juan is in the empty set.

STATISTICS

[696]   There are three kinds of lies: lies, damned lies, and statistics.

—Benjamin Disraeli (1804–1881) (also attributed to Mark Twain who, in *The Autobiography of Mark Twain*, actually quotes Disraeli.)

[697]   Statistical thinking will one day be as necessary for efficient citizenship as the ability to read and write.

—H. G. Wells (1866–1946)

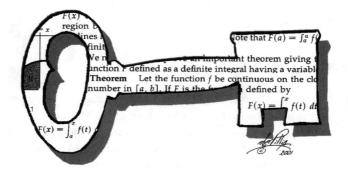

[698]

How well did H. G. Wells anticipate the future? Read on.

—JdP.

Algebra... once solely in place as the gatekeeper for higher math and the priesthood who gained access to it, now is the gatekeeper for citizenship; and people who don't have it are like the people who couldn't read or write in the Industrial Age.

—Robert P. Moses, Charles E. Cobb, Jr.
Radical Equations: Math Literacy and Civil Rights,
Beacon Press, 2001.

[699]   Statistics: the mathematical theory of ignorance.

—Morris Kline (1908–1992)
Mathematics in Western Culture, paperback,
Oxford Univ. Press, January 1965.

STYLE

[700]   You have erred perhaps in attempting to put colour and life into each of your statements instead of confining yourself to the task of placing upon record that severe reasoning from cause to effect which is really the only notable feature about the thing. You have degraded what should have been a course of lectures into a series of tales.

—Sir Arthur Conan Doyle (1859–1930)
spoken by Sherlock Holmes in *The Adventure of the Copper Beeches*

Note: See item **[162]** for more on cause (inductive reasoning) and effect (deductive reasoning).

SYMBOLS

[701]   Mathematics is often considered a difficult and mysterious science, because of the numerous symbols which it employs. Of course, nothing is more incomprehensible than symbolism which we do not understand.

Also symbolism, which we only partially understand and are unaccustomed to use, is difficult to follow. In exactly the same way, the technical terms of any profession or trade are incomprehensible to those who have never been trained to use them.

But this is not because they are difficult in themselves. On the contrary, they have invariably been introduced to make things easy.

—Alfred North Whitehead (1861–1947)

[702]   If a lunatic scribbles a jumble of mathematical symbols, it does not follow that the writing means anything merely because, to the inexpert eye, it is indistinguishable from higher mathematics.

—Eric Temple Bell (1883–1960)
In James R. Newman (ed.) *The World of Mathematics*,
New York: Simon and Schuster, 1956, pg. 308.

[703]   Such is the advantage of a well constructed language that its simplified notation often becomes the source of profound theories.

—Pierre-Simon de Laplace (1749–1827) quoted in "Mathematics in
the Modern World": Readings from the *Scientific American*, by
Morris Kline, W.H. Freeman & Co., San Francisco, 1968, pg. 795
[Chapter 33: "Determinants and Matrices" Epigraph.]

Note: One often cited example of symbolism that helped to advance theory is Wilhelm Gottfried Leibniz's fractional notation dg/dx for the derivative of the single-variable functions g with respect to variables x. The Leibniz symbol, dg/dx, as opposed to $g'(x)$ used by Isaac Newton, correctly suggests (but does not prove) the chain rule, namely, $df/dx = df/dg * dg/dx$. The notation of fractional cancellation only suggests the correct result but a rigorous proof must ultimately justify it.

[704] The training which mathematics gives in working with symbols is an excellent preparation for other sciences;... the world's work requires constant mastery of symbols.

—Jacob William Albert Young (1865–1948),
University of Chicago

Decimal Notation

[705] **Definition:** An **anagram** is a word or phrase created from another word or phrase made by transposing or rearranging the letters.

Example:

1. A Decimal Point == I'm a Dot in Place
2. Eleven plus two == Twelve plus one
3. The Morse code == Here come dots

[706] Before the introduction of the Arabic notation, multiplication was difficult, and the division even of integers called into play the highest mathematical faculties.

Probably nothing in the modern world could have more astonished a Greek mathematician than to learn that, under the influence of compulsory education, the whole population of Western Europe, from the highest to the lowest, could perform the operation of division for the largest numbers. This fact would have seemed to him a sheer impossibility.... Our modern power of easy reckoning with decimal fractions is the most miraculous result of a perfect notation.

—Alfred North Whitehead (1861–1947)
An Introduction to Mathematics
(New York and London, 1911), pg. 59.

[707]   I never could make out what those damned dots meant.

—Lord Randolph Churchill (1849–1895), British Conservative
politician. Referring to decimal points

Note: For Winston Churchill's mathematical epiphany, see item [465].

[708]   It is India that gave us the ingenious method of expressing all numbers by means of ten symbols, each symbol receiving a value of position as
well as an absolute value; a profound and important idea which appears so
simple to us now that we ignore its true merit. But its very simplicity and
the great ease which it has lent to all computations put our arithmetic in
the first rank of useful inventions; and we shall appreciate the grandeur of
this achievement the more when we remember that it escaped the genius of
Archimedes and Apollonius, two of the greatest men produced by antiquity.

—Pierre-Simon de Laplace (1749–1827)
In H. Eves, *Return to Mathematical Circles*,
Boston: Prindle, Weber and Schmidt, 1988.

[709]

Are numerals Arabic or Hindu? Since the ten digits,
*0,1,2,... 9, are commonly referred to as "Arabic
numerals" (as opposed to Roman numerals I, II, III,
IV,..., IX, X,...) why does Laplace, in item* [708], *credit
India with this notation? Howard Eves writes the
following:*

—JdP

The Hindu-Arabic number system is named after the Hindus, who may
have invented it, and after the Arabs, who transmitted it to western Europe.
The earliest preserved examples of our present number symbols are found
on some stone columns erected in India about 250 B.C. by King Aśoka.
Other early examples in India, if correctly interpreted, are found among
records cut about 100 B.C. on the walls of a cave in a hill near Poona and
in some inscriptions of about 200 A.D. carved in the caves at Nasik. These
early specimens contain no zero and do not employ positional notation. Positional value, however, and also a zero, must have been introduced in India

sometome before 800 A.D., for the Persian mathematician, al-Khowârizmi
desribes such a completed Hindu system in a book of 825 A.D.

—Howard Eves, *An Introduction to the History of Mathematics*,
section 1–9, "The Hindu-Arabic Numeral System."

SYSTEMS

[710]

> A system is a big black box
> of which we can't unlock the locks.
> And all we can find out about
> is what goes in and what comes out.
>
> Perceiving input-ouput pairs
> related by parameters,
> permits us sometimes to relate
> an input, output and a state.
>
> If these relations, good and stable
> then to predict we may be able.
> But if this fails us, heaven forbid
> we'll be compelled to force the lid.

—Kenneth Ewart Boulding (1910–1993), economist
Views on General Systems Theory, John Wiley & Sons, 1964.

T

TEACHING

[711]   If you ask me what accomplishment I'm most proud of, the answer would be all the young people I've trained over the years; that's more important than writing the first compiler.

—Grace Murray Hopper (1906–1992),
from *Computer Pioneers*, by J. A. N. Lee, Los Alamitos, CA:
IEEE Computer Society Press, pg. 386, 1991.

[712]

Grace Murray Hopper (1906–1992), was an early pioneer in computer programming. She joined the United States Naval Reserve during World War II and after USNR Midshipman's School, worked on the Bureau of Ordnance Computation Project at Harvard University. There she conducted award-winning work on Mark I, Mark II, and Mark III. Admiral Hopper's work with the Navy resulted in the development of programs and procedures for validating COBOL compilers.

—JdP

[713]   Do not then train boys to learning by force and hardness; but direct them to it by what amuses their minds.

—Plato (ca. 429–347 B.C.)

"Wow, I expected the calculator would help Bruce at school, but this?!"

[714] He who can, does—he who cannot, teaches.

—George Bernard Shaw (1856–1950)
Man and Superman (1903) Maxims for Revolutionists: Education

[715] He who can, does—he who cannot, criticizes.

—Anonymous

[716] The whole art of teaching is only the art of awakening the natural curiosity of young minds for the purpose of satisfying it afterwards.

—Anatole France (1844–1924) Nobel Laureate in Literature 1921,
born Jacques Anatole Thibault

[717] Some persons have contended that mathematics ought to be taught by making the illustrations obvious to the senses. Nothing can be more absurd or injurious: it ought to be our never-ceasing effort to make people think, not feel.

—A. T. Coleridge

Note: For concepts (dimensions) that go beyond the senses, see "Application 2: Breaking the Bonds of Three Dimensions," item **[617]**.

[718] Remember this, the rule for giving an extempore lecture is—let the mind rest from the subject entirely for an interval preceding the lecture, after the notes are prepared; the thoughts will ferment without your knowing it, and enter into new combinations; but if you keep the mind active upon the the subject up to the moment, the subject will not ferment but stupefy.

—Augustus De Morgan (1806–1871)

[719] To teach is to learn twice.

—Joseph Joubert (1754–1824), French philosopher

[720] A teacher affects eternity; he can never tell where his influence stops. A teacher is expected to teach truth, and may perhaps flatter himself that he does so, if he stops with the alphabet or the multiplication table, as a mother teaches truth by making her child eat with a spoon; but morals are quite another truth and philosophy is more complex still.

—Henry Adams (1838–1918)
The Education of Henry Adams, ch. 20.

[721]

No bubble is so irridescent or floats longer than that blown by the successful teacher.

—Sir William Osler (1849–1919) from a speech given in Glasgow, 4 Oct 1911, as quoted in: *Harvey Cushing, Life of Sir William Osler*, vol. 2, ch. 31, 1925.

[722]

> The decent docent doesn't doze:
> He teaches standing on his toes.
> His student dassn't doze—and does.
> And that's what teaching is and was.

—David McCord

Teaching That Fame Is a Worthy Achievement

[723] Our society has made fame an important desire, an occupation; whereas the really important people who are teaching our kids, are not seen as important because no one's writing about them.

> —Griffin Dunne (film maker) As quoted in the *Los Angeles Times*,
> July 14, 2001, pg. F1, from an article describing Hollywood films
> that present fame as a goal in itself.

 To see that society really does teach that fame is "an important desire," as Dunne claims in item **[723]** *above, look at the following items: The first (item* **[724]***) is a letter and reply from columnist "Dear Abby" and the second (item* **[725]***) is from a lead article in the Los Angeles Times.*

—JdP

[724]

DEAR ABBY: I am 12 years old and want to be a model, an actress and a singer. My mother is a housekeeper. My father is a supervisor in a factory. My aunts and uncles have similar jobs. Well, I don't want to do that. I want to be the next Daisy Fuentes or Cindy Crawford. I want people to know who I am. I don't want to be a "no one"—I want to be a "someone." My mother says not to get my hopes up because not too many people get famous. I say I don't care—I want to risk it. My parents think I might get hurt, but I know I am tough and can handle it. Abby, what do you think? Should I just forget about my dreams or "go for it"?
—WANTING TO BE A STAR in ELGIN, IL.

DEAR WANTING: Don't abandon your dreams. If you don't have dreams, how can they come true? Take classes in drama, dancing and singing. If they are not offered at your school, ask your parents how you

can earn extra money so you can take classes on Saturday. Don't neglect your schoolwork. You'll get even further if you have both brains and talent. There's no doubt in my mind you have the drive and determination to succeed. Good luck!
(c) August 23, 2001

[725]   [M]y daughter and I shared in our adoration of Aaliyah. All parents have that one celebrity we love as much as our children.... I sat on the edge of [my daughter's] bed to write these words, wanting to remember Aaliyah, the singer and budding actress. I didn't know her private life, but her image, her essence, warmed me.... That was good enough.

—Shonda Buchanan (lessons to her daughter after the accidental death of 22-year old rock star, Aaliyah Haughton) from a lead article in the *Los Angeles Times*, Sept. 2, 2001, Calendar section, pg. F1.

[726]

*The quote by **Griffin Dunne** (item [723]) is informed by Hollywood's depiction of fame, not as a result of any worthy achievement, but as a result without achievement. (See also [350].)*

But Dunne is speaking about more than teachers being under-valued and ignored by Hollywood. There are many cultural outlets expounding the message that fame is a goal in itself.

For example, in item [724], we read that 12-year old "Wanting" seeks fame and is inspired not by a Grace Hopper (for which see item [712]) or Jonas Salk, but by super-model, Cindy Crawford. If members of "Wanting's" family happen to be decent, loving, and hard-working, we would not know of it from this letter. However, we are told they are all "no ones" since they are shamefully unfamous. Dear Abby, who endorses "Wanting's" quest, does not dispute the implication that her family consists of "no ones," and, finally, offers no alternative to fame.

For another example, the Los Angeles Times *carried a lead article (item [725]) by Shonda Buchanan who concludes that a celebrity's "image" and "essence" (not the reality of the celebrity's character) warmed her. And that was "good enough."*

*Why should there be concern that self-esteem derives from
fame and not from any worthwhile achievement or skill?
(The key word, "worthwhile," is a discussion in itself.)
This question is relevant because in the classroom, there is
debate about a similar phenomenon—when a student's*
academic *self-esteem (i.e., self-esteem that usually results
from academic achievement) is granted by some educators
without the requirement of academic achievement.*

*Who would have guessed that this "results-without-
causes" philosophy is endorsed, alas, by none other than
Sherlock Holmes himself! (See item* [672].)

One of the allures of fame *is addressed by Henry
Kissinger in item* [727].

                                                              —JdP

[727]   The nice thing about being a celebrity is that when you bore people,
they think it's their fault.

                                                   —Henry A Kissinger
                                              Reader's Digest, April 1985

The R. L. Moore Method

[728]

Robert Lee Moore *(1882–1974) coined the term
"point-set topology." He is well-known for his unique
teaching method which, some would say, was less than
minimalist. Learning by the student and teaching of the
student was the responsibility of—the student. Moore's
pupils were obliged to avoid textbooks and all other
references in favor of "discovery." The result? After 64
years of teaching, Moore's 50 doctoral students include
three presidents of the AMS and five presidents of the
MAA.*

                                                              –JdP

Note: See also "The Greatest Math Teacher Ever" by Keith Devlin,
Devlin's Angle, June 1999, MAA online http://www.maa.org/devlin/ de-
vlin_6_99.html.

[729]   That student is best taught who is told the least.

—R. L. Moore (1882–1974) from *Comic sections: the book of
mathematical jokes, humour, wit, and wisdom,*
by Desmond MacHale, Dublin 1993.

[730]

Here's how we remember it.

John M. Worrell, Jr. was surely R. L. Moore's most fervent adherent. In
the movie, *Challenge in the Classroom*, Moore tells of a student, "Mr. W,"
who always left the room when others were presenting their results. Mr. W.
was John Worrell. He got an M.D. before earning his Ph.D. at Texas. Like
most of Moore's students, Worrell expressed himself very carefully. At a
Spring topology conference in Baton Rouge, Worrell was speaking about
Moore. He said, "I never knew his equal," paused to consider what he had
said, and added "Except for Moore himself."

—Tom Banchoff⋆ Brown University, and Bill Lindgren⋆,
Slippery Rock University private communication

THINKING

Note: For a discussion of the structure of thought, see also item **[20]**.

[731]   I think, therefore I am.

—René Descartes (1596–1650)
Le Discours de la Méthode, pt. 4, 1637.

[732]   I think, therefore I am single.

—Bumper sticker

[733]

When you believe in your own philosophy, there can be severe consequences, as the following story illustrates.

—*JdP*

One fine Spring afternoon, René Descartes was visiting his favorite country restaurant. After ordering a fine luncheon of fresh-caught trout from the local stream, he asked the waiter to recommend an appropriate white wine. "But monsieur Descartes," the waiter said apologetically, "unfortunately, we have no white wine in the cellar at all today. Would you care for a fine Burgundy instead?" "What!" Descartes replied vehemently. "You expect me to have trout with red wine? I THINK NOT!" At which point, he disappeared!

[734]   Never trust anything that can think for itself, if you can't see where it keeps its brain.

—by J.K. Rowling, as spoken by Harry Potter
in the book *Harry Potter and the Prisoner of Azkaban*, 1999.

[735]   It is a profoundly erroneous truism, repeated by all copy-books and by eminent people when they are making speeches, that we should cultivate the habit of thinking of what we are doing. The precise opposite is the case. Civilization advances by extending the number of important operations which we can perform without thinking about them.

—Alfred North Whitehead (1861–1947)
An Introduction to Mathematics,
Oxford University Press, 1992, pg. 42.

[736]   My mind rebels at stagnation. Give me problems, give me work, give me the most abstrusive cryptogram, or the most intricate analysis, and I am in my own proper atmosphere.

—Sir Arthur Conan Doyle (1859–1930)
spoken by Sherlock Holmes in *The Sign of the Four*

[737]

> Cannon-balls may aid the truth,
> But thought's a weapon stronger.
> We'll win our battles by its aid,
> Wait a little longer.

—Charles Mackay (1814–1889)
There's a good time coming, boys, 1846 poem.

[738]   Thought is only a flash between two long nights, but this flash is everything.

—Henri Poincaré (1854–1912)
Quoted in James R. Newman,
The World of Mathematics, New York, 1956.

TOPOLOGY

Note: See also R. L. Moore Method, item **[728]**.

[739]   A topologist is one who doesn't know the difference between a doughnut and a coffee cup.

—John Kelley (1916–1999)

TRANSCENDENTAL NUMBERS

[740]

Definition (1): Algebraic numbers are the real roots of single variable polynomials whose coefficients are integers.

Definition (2): Transcendental numbers are all the other real numbers, that is, numbers that are not algebraic.

Fact (3): If, in Definition (1), you replace integer coefficients with more general algebraic number coefficieants, then the roots are still algebraic.

Fact (4): Algebraic numbers are well-structured in that they form a field. That is, sums, differences, products, and quotients of algebraic numbers (except for division by 0) are still algebraic numbers.

Fact (5): All transcendental numbers are irrational numbers. But, the converse is not true; there are some irrational numbers (e.g., $\sqrt{2}$) that are not transcendental.

Fact (6): Here are two specific transcendental numbers:

e, the base of the natural logarithm. Transcendence of e was proved by Charles Hermite (1822–1901) in 1873, and

π, *the ratio of a circle's circumference to its diameter in a plane. Transcendence of π was proved by Ferdinand Lindemann (1852–1939) in 1882. (For more on π, see items [522]–[529].)*

About pi. Since pi is transcendental, it is impossible to "square the circle," that is, using a compass and straightedge in accordance with the ancient Greek rules for geometric constructions, to draw to a square with exactly the same area of a given circle.

For proofs, see Transcendental Number Theory, *by A. Baker, Cambridge University Press 1975, Chapter 1.*

—JdP

[741]

About the structure of transcendental numbers. *In contrast to the closed field structure of algebraic numbers described in* **[740]**, *Fact (4), it is not always known whether sums and products of transcendental numbers are transcendental. For example,*

Fact (7): *At least one of the two numbers,*

$$(\pi + e) \quad and \quad (\pi \times e),$$

is transcendental—they can't both be algebraic—yet, it is not known which one it is.

An easy proof: *Look at this!*

[742] $$Y^2 - (\pi + e)Y + (\pi \times e) = (Y - \pi) \times (Y - e) = 0.$$

Now assume the opposite of what we are trying to show, namely, that both coefficients of the left-hand polynomial in **[742]** *are algebraic. (See items* **[598]**, *for more on proof by contradiction.) Item* **[740]**, *Fact (3) above says the roots, $Y = \pi$ and $Y = e$, (see right-hand side of* **[742]***) must be algebraic while, at the same time,* **[740]** *Fact (6) says π and e are both* transcendental *(not algebraic). This contradiction forces us to believe that at least one of the coefficients, $(\pi + e)$ and $(\pi \times e)$, is transcendental—they can't both be algebraic.*

—JdP

Note: This discussion owes much to Jan-Hendrik Evertse of the Universiteit Leiden, the Netherlands.

[743] Transcendental numbers occupy a position in the field of real or complex numbers much like that of insects in the kingdom of animals. Everybody knows they are, by a large margin, the most abundant class, but few know more than one or two of them by name.

—Donald R. Newman

TRANSFINITE NUMBERS

Note: See also "Infinity," item [254].

[744]

> Georg Cantor took sets and his view of 'em
> Changed deeply the way we would look at 'em.
> > He showed that infinity
> > Has an affinity
> To break up and spawn more than one of 'em.

—John de Pillis

U

UNITS

[745]

$$10^6 \text{ phones} = 1 \text{ megaphone}$$

$$2^3 \text{ puss} = 1 \text{ octopus}$$

$$10 \text{ carts} = 1 \text{ Descartes}$$

$$\text{mate} = 1 \text{ decimate}$$

$$10^{-1} \text{ dents} = 1 \text{ decadent}$$

$$2,000 \text{ mockingbirds} = \text{two kilomockingbird}$$

UNIVERSITY DENIZENS

[746]

 It was frustrating for Tom Lehrer. *His original lyrics for "The Professor's Song," published in 1975 in the American Mathematical Monthly, carried an unfortunate misprint that altered the impact of one particular line. A saw-tooth pattern, something like the ascii sequence,*

*was printed instead of a handwritten, cursive "design"
which should have appeared. Lehrer scribbled out the
design he had originally intended and this scribbling is
used in the cartoon of item* [747].

—*JdP*

[747]

The Professor's Song

If you give me your attention, I will tell you what I am.
I'm a brilliant math'matician—also something of a ham.
I have tried for numerous degrees, in fact I've one of each;
Of course that makes me eminently qualified to teach.
I understand the subject matter thoroughly, it's true,
And I can't see why it isn't all as obvious to you.
Each lecture is a masterpiece, meticulously planned,
Yet everybody tells me that I'm hard to understand,
    And I can't think why,

My diagrams are models of true art, you must agree,
And my handwriting is famous for its legibility.
Take a word like "minimum" (to choose a random word),
For anyone to say he cannot read that, is absurd.
The anecdotes I tell get more amusing every year,
Though frankly, what they go to prove is sometimes less than clear,
And all my explanations are quite lucid, I am sure,
Yet everybody tells me that my lectures are obscure,
    And I can't think why.

Consider, for example, just the force of gravity:
It's inversely proportional to something—let me see—
It's r^3—no, r^2—no, it's just r, I'll bet—
The sign in front is plus—or is it minus, I forget—
Well, anyway, there is a force, of that there is no doubt.
All these formulas are trivial if you only think them out.
Yet students tell me, "I have memorized the whole year through
Ev'rything you've told us, but the problems I can't do."
    And I can't think why!

—Tom Lehrer—(Tune: *If You Give Me Your Attention* from
Princess Ida, William S. Gilbert and Sir Arthur Sullivan.)
The American Mathematical Monthly, vol. 81, 1974, pg. 745.

[748]   No discipline is ever requisite to force attendance upon lectures
which are really worth the attending.

—Adam Smith (1723–1790)

[749]   A professor is one who can speak on any subject—for precisely fifty
minutes.

—Norbert Wiener (1894–1964)

[750] **The Professor**   A professor was being interviewed by a mathematics department at another university for a well-paying tenured position in
algebra. Consistent with his overbearing nature, the professor wanted to be
sure he would not encounter factions that might interfere with his personal
ambition to dominate the department. He therefore hired a private eye to
make a full investigation of the department.

When the investigator submitted his report, it read: "The Mathematics
Department at the University has an excellent reputation. Due to the their
well-structured hiring program, the Department has never suffered from internal strife, petty bickering, or factionalism. There is a high degree of de-
partmental collegiality and harmony. Recently, however, the Department
has been interviewing an extremely irascible and disagreeable candidate for
the algebra position."

—as told by John de Pillis

[751]   This is a chain letter. If you want to get ahead in the academic com-
munity follow the directions given herein.

A Chain Letter for Survival

1. Add the three names listed below as co-authors of two of your current publications.
2. Remove the last name on the list below and place your name in the first position.
3. Send copies of the revised letter to five of your colleagues within the next week.

In just a few weeks you will be listed as co-author of hundreds of publications. This will certainly result in increased prestige, international recognition, large raises, and early promotions.

Do not break the chain! A. N. Onymous broke the chain and he was dismissed from a prominent university for lack of publications. No one has heard of Dr. Onymous since. On the other hand, Carlitz and Erdös are doing quite well.

Three Name List

1. P. O. Perish
2. H. Petard
3. N. Bourbaki

Sincerely,
P. O. Perish

Note: See also George Bush, item **[110]**, for his brain proclamation. For Abraham Lincoln on toggling and other matters, see items **[274]**–**[276]**. For more on George Washington and his comments on gambling, see items **[768]**–**[771]**.

[752]   I find the three major problems on a campus are sex for the students, athletics for the alumni, and parking for the faculty.

> —Clark Kerr quoted in *An Irreverent and Thoroughly Incomplete Social History of Practically Everything*, by Frank Muir, Stein and Day, New York, 1976.

[753] The University President   To improve public relations, the university president initiated a series of non-technical public lectures to be given by members of the faculty. At the first lecture, the president introduced the evening's speaker, a mathematician, who chose the topic, "The Marvel of Zero."

The mathematician did a wonderful job explaining in lay terms how zero is both a number (as $0 + 3 = 3$) and a placeholder (where 207 is distinguished from 27). He explored the history of place holding notation going back to ancient Babylon. The professor noted that the invention of zero as a number was a great intellectual leap since, when seen as an additive "identity," spurred the development of group theory.

At the public reception following, a man from the audience approached the president and asked him for his reaction to the lecture.

"The mathematics professor has a real lesson for all other departments of this university," the president said.

"Oh, I see," said the man. "You mean the mathematician communicated so clearly across disciplines that he broke through the walls of overspecialization."

"Not exactly," the president replied. "The departments should learn that when they ask me for resources, they should be as thankful as the mathematicians are for nothing."

> —as told by John de Pillis

[754] The Dean   With only days remaining until the start of the semester, the dean of the science faculty was desperate to fill a vacancy for a lecturer of a popular differential equations course that already had an enrollment of 150 students.

After a day of frustratingly futile telephone calls and interviews, the dean went to the local pub to relax over a beer. A dog behind the bar took his order and put his beer in front of him, took the money to the cash register, rang up the bill a and returned with exact change.

Wiping the bar with a rag, the dog said, sympathetically, "You look a bit stressed." The dean replied, "You bet I am. At the very last minute, I have to find a qualified lecturer to teach a course in ordinary differential equations."

"Ah," the dog said wryly, "isn't it funny how they call those equations 'ordinary?' Yet with their many applications, not to mention the power that numerical methods provide, they really should call them 'extra-ordinary' differential equations. Don't you agree?"

The dean was stunned. As the conversation continued, the dean saw that the dog's knowledge was profound and his explanations were succinct, intuitive, and clear.

"Say, I could really use this dog to teach that course," the dean said to the owner. "I want to hire him as a lecturer as soon as possible!" "Nah, he wouldn't be any good to you as a lecturer," the owner said, "his handwriting is terrible!"

—as told by John de Pillis

[755] **Chairman**   At a department meeting, the mathematics chair was trying to encourage the faculty to improve their teaching. As the list of student complaints made clear, members of the Department needed to be more sensitive to student's needs with respect to office hours, quality of lectures, and

homework. As the criticisms were discussed, the chair noticed that each faculty member looked at the others as if in reproach. Does each faculty member think the fault is with someone else, he thought?

To find out, the chair said, "Isn't it paradoxical? Those faculty who are the most capable teachers happen to be the ones who can't even tie their own ties straight." The department chair knew he had a problem when every faculty member suddenly reached to straighten his tie.

—as told by John de Pillis

[756] The Graduate Student   The graduate student visited the University Psychological and Counseling Center.

"I really need help," he said to the counselor, "Every chance I get, I'm out partying or wasting time watching TV. I'm ignoring my responsibilities. It's terrible—I feel so guilty every time I find myself goofing off and giving in to these urges."

"So you want me to help you strengthen your resolve," the counselor said sympathetically.

"Heck, no," the student replied, "I want you to weaken my conscience!"

—as told by John de Pillis

Graduate Student

USEFULNESS

[757]   Newborn Babies: When [Michael] Faraday (1791–1867) first made public his remarkable discovery that a changing magnetic flux produces an emf, he was asked (as anyone is asked when he discovers a new fact of nature), "What is the use of it?" All he had found was the oddity that a tiny current was produced when he moved a wire near a magnet. Of what possible "use" could that be? His answer was: "What is the use of a newborn baby?"

—Richard P. Feynman, (1918–1988)
The Feynman Lectures on Physics,
Volume II, Section 16-4, Electrical Technology.

[758]

Mathematics? How else could a single lady determine the direct correlation between her suitor's IQ and the carat weight of the diamond he presents to her? How else could the (eligible) suitor compute the statistical average of getting a "yes"? Mathematics is absolutely romantic!

—Gretchen de Pillis

[759]   Useful Military Terminology   To avoid *corporal* punishment, might not a *major* function of the *kernel* function be replaced by a *general* function? In *private*, of course. Indeed, an *admiral*-able concept.

—Allan Edelson*, University of California, Davis

V

VALENTINES

[760]

Take one over x and its integral
From naught to a point serendipital.
    My love is thus measured
    For you, my sweet treasure
Clearly quite large and infinital

—John de Pillis

[761]

My valentine I send to thee
With love and osculation.
Dear love let's have no more harmonic separation
And I wait thy reciprocation.

—Leo Moser The American Mathematical Monthly,
vol. 80, no. 8, 1973, pg. 902.

[762]

You disintegrate my differential,
    you dislocate my focus.
My pulse goes up like an exponential
    whenever you cross my locus.
Without you, sets are null and void—
    so won't you be my cardioid?

—Katherine O'Brien Mathematics Teacher,
vol. 58, no. 6, pg. 537, 1965.

VIRUS

[763]   **The computer virus and the microbial virus**   Indeed, the microbial biosphere can be thought of as a World Wide Web of informational exchange, with DNA serving as the packets of data going every which way. The analogy isn't entirely superficial. Many viruses can integrate (download) their own DNA into host genomes, which subsequently can be copied and passed on: Hundreds of segments of human DNA originated from historical encounters with retroviruses whose genetic information became integrated into our own genomes.

What makes microbial evolution particularly intriguing, and worrisome, is a combination of vast populations and intense fluctuations in those populations. It's a formula for top-speed evolution. Microbial populations may fluctuate by factors of 10 billion on a daily cycle as they move between hosts, or as they encounter antibiotics, antibodies, or other natural hazards. A simple comparison of the pace of evolution between microbes and their multicellular hosts suggests a million fold or billion fold advantage to the microbe. A year in the life of bacteria would easily match the span of mammalian evolution!

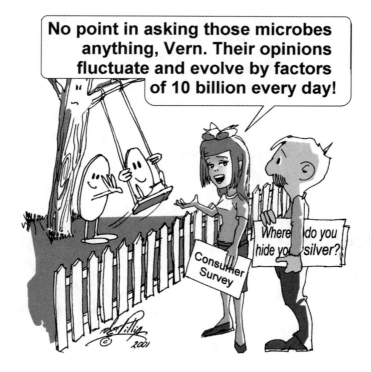

(added for this edition) God help us, someday the computer viruses may also be designed to "evolve." Although less likely, such evolution might even occur by happenstance.

—Joshua Lederberg*,
Nobel Laureate, Physiology & Medicine 1958
Science 2000 April 14; 288:291
(in *Pathways of Discovery*)

[764]

Joshua Lederberg was 33 *when he was awarded the Nobel Prize (shared with George Wells Beadle and Edward Lawrie Tatum) in Physiology or Medicine in 1958. From the presentation speech by Professor T. Caspersson,*

> *Particularly important is [Lederberg's] discovery that sexual fertilization is not the only process leading to recombination of characters in bacteria. Bits of genetic material can, if they are introduced into the bacterial body, become part of the genetic material of the bacterial cell and thus change its constitution. This is usually termed "transduction," and it is the first example demonstrating that it is possible experimentally to manipulate an organism's genetic material and to introduce new genes into it...*

–JdP

VOCABULARY

[765]

Ambiguous case — politician
Bit — information from the horse's mouth
Directrix — female director
Epsilon — married to Delta

Falling body — whodunit

Group theory — encounter dynamics

Im grossen problem — obesity

Joint variation — arthritis

Knots — traffic jam

Linear operator — directory assistance

Measure of dispersion — litterbug

Necessary and sufficient — money

One-to-one correspondence — love letters

Plane, intercept form — hijacker

Quantifier, universal — income tax

Related rates — cops & robbers

Significant digit — hitchhiker

Transcendental function — guru

Upper bound — price controls

Volume — rock 'n' roll

Work (ft-lb) — overweight jogger

$X \cup Y$ — togetherness

—Katherine O'Brien *Mathematics Magazine*,
vol. 52, no. 5, pg. 291, 1979.

VON NEUMANN AND THE FLY

[766]   This story about von Neumann involves an old problem about two bicyclists and a fly:

Two bicyclists, initially 60 miles apart, travel towards each other at 10 miles per hour. At the moment they start, a fly takes off from the front rim of one bike and travels towards the second bike at 20 mph. After reaching the second bike, the fly instantly turns around and flies towards the first bike at 20 mph, continuing this procedure until the bikes meet. How far does the fly travel?

The easy way to solve this problem is to note that the bikes meet in three hours, thus the fly travels a total of $3 \times 20 = 60$ miles.

A harder way is to add up individual distances travelled by the fly between successive contacts with the bikes. This yields the geometric series

$$40 \times \left(1 + \tfrac{1}{3} + \left(\tfrac{1}{3}\right)^2 + \dots\right) = \frac{40}{1 - \tfrac{1}{3}} = 60.$$

Now back to the story. Von Neumann had a well-deserved reputation for solving problems and performing difficult calculations in his head, and he loved a challenge. One day at a cocktail party, a young colleague posed the fly problem to von Neumann. He solved it in just 8 seconds. The young colleague then told von Neumann he had presented the same problem to a physicist who solved it in 6 seconds. Neumann was somewhat astonished and said "Impossible! No physicist can sum a geometric series in 6 seconds."

Note: This story has appeared in various publications and the numbers vary in almost all of them. One version can be found in, *John Von Neumann: The Scientific Genius Who Pioneered the Modern Computer, Game Theory, Nuclear Deterrence, and Much More*, by Norman Macrae, Amer. Math. Soc. 1999.

W

WASHINGTON

[767]

Everyone knows that Washington was born in the year "Integer part of $(1,000\sqrt{3})$."

[768]   On Games of Chance: ...avoid Gaming. This is a vice which is productive of every possible evil, equally injurious to the morals and health of its votaries. It is the child of Avarice, the brother of Inequality, and father of Mischief. It has been the ruin of many worthy familys [sic]; the loss of many a man's honor; and the cause of Suicide. To all those who enter the list, it is equally fascinating; the Successful gamester, in hopes of retrieving past misfortunes, goes on from bad to worse; till grown desparate, he pushes at every thing; and loses his all...few gain by this abominable practice (the profit, if any, being diffused) while thousands are injured.

—George Washington (1732–1799)
From a letter to Bushrod Washington, Jan. 15, 1783, *Writings*
Vol. 26, pg. 40, as cited in *Maxims of George Washington*,
The Mount Vernon Ladies Association (no kidding!) 1989, pg. 154.

Note: See also **[565]** for comments on chance and determinism.

[769]

Washington was so popular a figure after the Revolutionary War that a number of Army officers asked him to be King or Emperor of America. Since 29 B.C., when Augustus Caesar (63 B.C.–14 A.D.) was made Emperor of Rome, the world had only known unrepresentative, non-consensual governments ruled by kings or emperors. (Before Augustus, there was the Roman Republic, which, with its Consuls, Senators, Praetors, Lictors, Quaestors, did not at all resemble a democracy or a republic except in name.) By refusing to be King, Washington deliberately interrupted the momentum of nearly 2,000 years of dictatorships. His refusal to be King prevented the officers from carrying out a planned military takeover of the country in 1783. Later that year, Washington and his officers voluntarily resigned their military positions.

—JdP

Note: See items **[103]** and **[104]** for details of Aristotle's view of democracy and governemnt.

[770]

As Robert Frost would note, "George Washington was one of the few in the whole history of the world who was not carried away by power."

—*JdP*

[771]

Washington was the first man to sign the U.S Constitution as a delegate from Virginia. A lesser accomplishment was that he was responsible for introducing the mule to America in the 1780's.

—*JdP*

WIENER

[772]

There once was a Norbert named Wiener
Whose mind couldn't be keener
    But he'd chant and recite
    Verses so trite
That we wished he'd sing less and obscener.

—Leo Moser *The American Mathematical Monthly,*
vol. 80, no. 8, 1973, pg. 902.

X-Y-Z

X

[773] To speak algebraically, Mr. M. is execrable, but Mr. G. is $(x + 1)$ecrable.

—Edgar Allen Poe (1809–1849)

"HEY, WAIT A MINUTE! JUST YESTERDAY, SHE SAID X WAS EQUAL TO FIVE!"

[774]

When Zeno was still a young man
Impressed with the way turtles ran,
   He challenged Achilles
   And some say still he's
Not certain which one's in the van.

—Paul Ritger

ZERO

[775] A zero of order zero is a regular point at which the function is not zero.

—From a book on complex variables.

[776]

Both minus and plus, I'm the same.
There's no one else can make that claim.
   Add, subtract me without alarm,
   Or multiply—it does no harm.
But—
Mind the rule, be not a hero:
"Thou shalt not divide by zero!"

—Nick J. Rose

[777]

Formula for Notoriety

If you want to be famous like Miro,
Write a book on division by zero,
 Or give a long lecture
 On Goldbach's conjecture:
You'll end up a bit of a hero.

—John McClellan, *The Mathematics Teacher*,
vol. 59, no. 7, pg. 655, 1966

Note: For the statement of the **Goldbach Conjecture**, see item [331].

Note: The Starlight Café

John de Pillis, *Starlight Café Conversations: An Illustrated Dictionary from Table Seven* (to appear)

At the Starlight Café, the following group meets regularly at Table Seven for conversation and discussion.

- *Anvil Willie*, an enigmatic man of wide and varied experience.

- *Cordelia Morrow Layton*, a trusting, good-hearted person who shares her feelings freely.

- *Mona "Hutch" Hutchinson*, a university student and part-time taxi-cab driver. As Hutch is curious and friendly, most of our "outside" specialty guests come through her invitation.

- *JdP*, I record the conversations, organize them, and add sketches.

Index of Authors and Topics

LEGEND

Note: All quotes herein, attributable to the cited author and appearing for the first time in print, are indicated by the symbol *.

Note: Conversation Starters are denoted by square brackets, as [123]. Page numbers appear as unbracketed numerals, as 123.